Plant Biotechnology in Agriculture

K. Lindsey and M.G.K. Jones

JOHN WILEY & SONS
Chichester · New York · Brisbane · Toronto · Singapore

First published 1989 by Open University Press

Reprinted April1992, January 1995 by John Wiley & Sons Ltd,
Baffins Lane, Chichester, West Sussex PO19 1UD, England

Other Wiley Editorial Offices

John Wiley & Sons, Inc., 605 Third Avenue,
New York, NY 10158-0012, USA

Jacaranda Wiley Ltd, G.P.O. Box 859, Brisbane,
Queensland 4001, Australia

John Wiley & Sons (Canada) Ltd, 22 Worcester Road,
Rexdale, Ontario M9W 1LI, Canada

John Wiley & Sons (SEA) Pte Ltd, 37 Jalan Pemimpin # 05-04,
Block B, Union Industrial Building, Singapore 2057

British Library Cataloguing in Publication Data

Lindsey, Keith
Plant biotechnology in agriculture
1. Plants. Biotechnology
I. Title II. Jones, M. G. K. III. Series
660′.6

ISBN 0 471 93238 8

Typeset by Vision Typesetting, Manchester
Printed and bound in Great Britain by
Antony Rowe Ltd, Chippenham, Wiltshire

The Biotechnology Series

This series is designed to give undergraduates and practising scientists access to the many related disciplines in this fast developing area. It provides understanding both of the basic principles and of the industrial applications of biotechnology. By covering individual subjects in separate volumes a thorough and straightforward introduction to each field is provided for people of differing backgrounds.

Published Titles

Biotechnology: The Biological Principles M. D. Trevan, S. Boffey, K. H. Goulding and P. Stanbury
0 471 93279 5 CL 272pp 1987
0 471 93278 7 PR 272pp 1987

Fermentation Kinetics and Modelling C. G. Sinclair and B. Kristiansen (Ed. J. D. Bu'lock)
0 471 93208 6 CL 128pp 1987
0 471 93207 8 PR 128pp 1987

Enzyme Technology P. Gacesa and J. Hubble
0 471 93223 X CL 176pp 1987
0 471 93222 1 PR 176pp 1987

Animal Cell Technology: Principles and Products M. Butler
0 471 93210 8 CL 192pp 1987
0 471 93209 4 PR 192pp 1987

Fermentation Biotechnology: Principles, Processes and Products O. P. Ward
0 471 93286 8 CL 240pp 1989
0 471 93285 X PR 240pp 1989

Genetic Transformation in Plants R. Walden
0 471 93282 5 PR 160pp 1988

Plant Biotechnology and Agriculture K. Lindsey and M. Jones
0 471 93239 6 CL 256pp 1989
0 471 93238 8 PR 256pp 1989

Biosensors E. Hall
0 471 93226 4 CL 368pp 1990
0 471 93224 8 PR 368pp 1990

Biotechnology of Biomass Conversion M. Wayman and S. R. Parekh
0 471 93288 4 CL 256pp 1990
0 471 93287 6 PR 256pp 1990

Biotechnology in the Food Industry M. P. Tombs
0 471 93277 9 CL 208pp 1990
0 471 93276 0 PR 208pp 1990

Bioelectronics S. Bone and B. Zaba
0 471 93296 5 CL 160pp 1992

Series Editors

Preface and Acknowledgements

Agriculture is an ancient practice, which, in the last century or so, and in particular since the ideas of Darwin and Mendel were published, has been moulded into a science through the developments of modern genetics and biochemistry. Within the last twenty years advances in our understanding of plant cell and molecular biology have provided the basis for a new crop of techniques which constitute plant biotechnology. This is an umbrella term which covers the concepts and methodologies of plant tissue culture and genetic manipulation, and it is the aim of this book to introduce these concepts (and to a lesser extent their technical details), and in particular to explore their application to agricultural problems.

The whole field of plant biotechnology is expanding at an enormous rate, and it is not possible to cover all aspects in great detail. A related danger in writing a book of this kind is that it may rapidly become dated. Therefore, it has been our intention to introduce the basic scientific ideas which have formed the keystones in plant biotechnological research, to describe how these have led to major advances in the application of this work, and to indicate some of the more important aims and objectives of future research, as we see them.

Tissue culture and genetic manipulation techniques will not replace conventional plant breeding, but can be considered as valuable tools to investigate plant gene structure and expression and the development of the organism, and also to provide genetically-modified material which can be integrated into an established breeding programme. We have therefore tried to illustrate how the techniques can be exploited to these ends, with examples of the use of transformation systems to study chimaeric gene expression, of approaches to isolate specific genes of interest, and of tissue culture and regeneration techniques to produce breeding material or valuable plant metabolites. Because we are at a relatively early stage in the evolution of these strategies, and in particular because very few agronomically important genes have been isolated to date, much of what we have written is speculative, particularly in relation to the manipulation of plant development. We hope, however, that we have indicated the strengths and weaknesses of current techniques, and have highlighted some hurdles to be overcome, so that the reader can form his or her own opinion of the ways in which specific agricultural problems may be tackled.

We feel that this volume will be of interest to advanced undergraduate and post-graduate students of plant science or biotechnology, and to working scientists specialized in a different discipline. The reader is assumed to have a basic knowledge of plant biochemistry, but the jargon of molecular biology, which can be as much of a barrier to their assimilation as the complexities of the techniques themselves, is kept to a minimum, and a glossary of terms is provided. Specific references to research papers are also minimized, in keeping with the style of the series, and so the authors of some research may not be cited directly. Access to the literature is, however, possible through the articles quoted.

Though the responsibility for the accuracy and style of the text is solely ours, we would like to acknowledge with thanks the efforts of Dr M. Kreis, Prof. C.J. Leaver FRS, Dr P.R. Shewry, Dr J.F. Topping and Dr P. Wilkins for critically reading parts of the manuscript, Mrs Jeanne Hutchins for typing parts of the text, and to numerous colleagues at Rothamsted for varied and valuable discussions. Finally, we are grateful to Richard Baggaley of Open University Press for his help and encouragement during the preparation of this volume.

K. Lindsey
M.G.K. Jones
Leicester Biocentre,
University of Leicester and
Rothamsted Experimental Station,
Harpenden

Chapter 1

Established Plant Breeding Techniques – Successes and Limitations

Introduction

Agriculture is the most important activity carried out by mankind, and one of the first steps in agriculture was the domestication of specific plants as crops. The process of domestication involves recognition that a particular plant species has something to offer. As a result, selection of better plants has occurred over thousands of years, and has been a continuous process carried out both consciously and unconsciously.

The aims for selection of improved crop plants have remained relatively constant, with crop plants being selected for ease of cultivation, greater yield, appropriate quality and resistance to pests, diseases and environmental stress. However, the current technology of plant breeding has developed more recently, over the last century. It is founded on the greater understanding of the genetic basis of agricultural traits, and more particularly, by application of Mendel's classical genetic principles. However, in order to achieve breeding aims, success in breeding also depends on an understanding of plant physiology, biochemistry and pathology. It is also necessary to take into account the final usage of the harvested part of the crop. Apart from direct selection of the best plants from existing populations, plant breeding at the simplest level involves the deliberate hybridization of two parental plant genotypes, selected for specific traits, followed by selection of novel recombinant genotypes combining the specific traits of interest.

Table 1.1 Some inbreeding and outbreeding species

Self-pollinated species	*Cross-pollinated species*
Wheat	Maize
Barley	Rye
Oats	Brassicas: cabbage etc.
Rice	Sunflower
Tomato	Rye grasses
Egg plant (aubergine)	Clovers
Flax	Potato
Peach	Beet
Peanut	Radish
Tobacco	Carrot
Cotton	Mango
Soybean	Cucurbits: marrow, squash etc.
Lentil	Yams
Pea	Rubber
Cowpea	Banana
Coffee (*Coffea*)	Coconut
Pepper (*Capsicum*)	Pineapple
Safflower	
Lettuce	*Vegetatively propagated species*
	Potato
	Sugar-cane
	Orchard and soft fruits

The most important factor that determines the breeding approach to be taken, is the breeding system of the crop. Plants may either be mainly self-fertilized, in which case they are referred to as inbreeding species, or there may be inherent barriers to self-compatibility such that an egg cell is normally only capable of being fertilized by pollen of a different genotype of that species. In this case they are referred to as outbreeding species. A list of some inbreeding and outbreeding species is given in Table 1.1. Some crops are propagated vegetatively, such as sugar-cane, potato, orchard crops and soft fruits. Most of these are outbreeders, in which natural or artificial methods of clonal reproduction lead to bypassing of sexual reproduction.

There are four genetic categories of crop cultivars. These are (1) inbred lines, (2) hybrids, (3) clones and (4) populations of plants. The reasons for these categories are discussed below, and depend mainly on the mode of pollination of the crop plants concerned.

Single Gene and Polygenic Traits

Agronomic characters that are determined by a few genes with clear effects are much simpler and more effective to select than traits determined by the additive effects of many genes each of limited individual effect. Plant breeders have therefore tended to utilize single-gene variants in preference to polygenic variation. However, analysis, interpretation and prediction of the effects of genes with limited individual effect (polygenes) can be carried out. This requires a more rigorous approach, based as much on statistical as genetical analysis, and this approach has been referred to as 'biometrical genetics'. As the number of desired genes (n) selected for in a crop increases, the number of plants that need to be evaluated to obtain the superior genotypes possessing all the desired combination of genes increases by a factor of 2^n (assuming no linkage between loci). This figure is based on two alleles per locus, and is an underestimate since multiple alleles occur at some loci. Also, some useful loci are linked, and dominance and interaction between loci can occur along with interactions of genotype with environment, which further increases the ideal number of progeny to be evaluated. Thus plant breeders need in general to work with large populations of plants, and have traditionally looked for what they believe are 'ideal plant types' (or 'ideotypes') in addition to selection for yield and other screenable characters.

General Pattern of Plant Breeding

The general system of plant breeding is given in Fig. 1.1 (from Simmonds, 1979), and indicates the range of factors that a breeder must consider and the decision that he must take. The breeder must decide what parental plants to use, how to combine them, what to select for, how to select, what to discard and what to keep. Some of these aspects are considered further. A notable feature of most plant breeding programmes is the time-scale, the period from the initial cross to marketing a variety, often 15 years or longer.

Inbreeders and Pure Line Varieties

The genetic effect of continued self-fertilization in self-pollinated species is to separate out dominant and recessive genes, dominant genes being represented by capital letters (A) and recessive genes by lower case (a). Following Mendel's rules, if the heterozygote Aa is selfed, the three genotypes produced are AA, Aa and aa, in the ratio 1 : 2 : 1 in the next or F_1 generation. The 'genotype' is the genetic constitution of an individual, and two plants that may look identical can have different genetic constitutions (e.g. AA, Aa), because of the dominance of one allele over another. The 'phenotype' of an organism is the appearance and function of an organism as a result of its genotype and its environment. When the genotypes AA and aa are selfed, then they breed true, whereas selfing the Aa genotype, which is phenotypically indistinguishable from the AA genotype, results in the same segregation of progeny as before. The heterozygosity of the population is thus reduced by half in each generation, and after six or seven generations of selfing, the population will consist almost completely of the two genotypes AA and aa. In this way homozygosity can be achieved by selfing to give pure breeding lines.

If the array of genes present in an inbreeding species is considered, after selfing

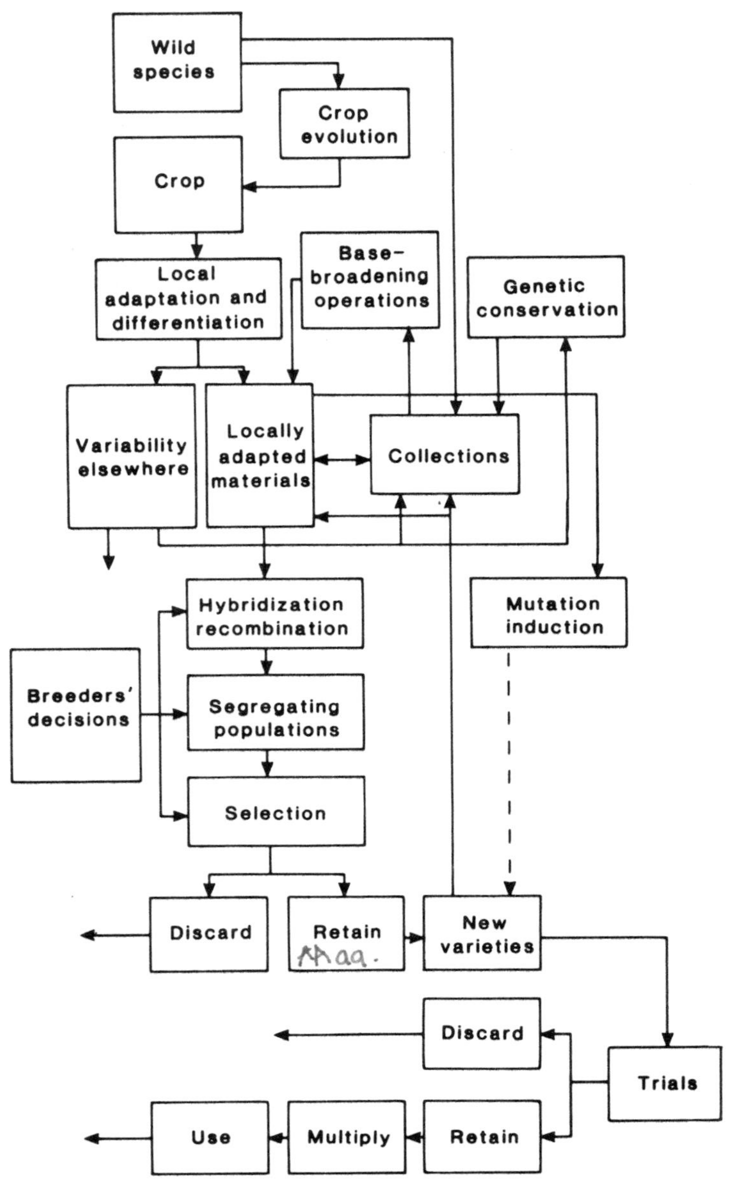

Fig. 1.1 General pattern of events involved in plant breeding. Breeders work with a gene pool of locally-adapted parental plants, intermittently supplemented from elsewhere to maintain an adequate genetic base. Breeders must decide what parents to include, how to combine them, what to select for, how to select, what to discard and what to keep. New varieties submitted by the breeder for practical exploitation are subject to further decisions on

the population will consist of many different genetically homozygous lines. There will be genetic differences between the lines, but not within them. Both major genes with clear phenotypic effects and polygenes with additive effects will behave in this manner. As a result of selection for characters by continued selfing, pure lines are thus obtained. A desired character selected in this way is said to be 'fixed', and the character breeds true. In a plant-breeding programme, it is clearly necessary to select promising lines and discard the rest, for the number of different lines that could be produced is virtually infinite.

From this it can be seen that no new variability is introduced during the selection scheme, and the only genetic variability is present in the initial plant used. The potential for obtaining superior genotypes is therefore increased by first generating genetic variability by making artificial crosses, by hand, between two selected parents which exhibit potentially valuable agronomic properties that it would be desirable to combine.

Hexaploid wheat is an inbreeding species that can be treated in this manner, and a typical wheat-breeding scheme is shown diagrammatically in Fig. 1.2. In this example taken from the former Plant Breeding Institute at Cambridge, about 1100 crosses are made each year and 1800 plants from each cross are grown in the second generation (F_2). At this stage, about two million plants must be examined in the field for disease resistance and agronomic appearance, especially height and ear type. At this stage it is also possible to select on a single plant basis for grain appearance and some quality traits. Progenies from about 50 000 plants from the F_2 generation are grown separately in rows derived from single ears in the F_3 generation, with more rigorous selection. Selected lines are grown again as ear rows in the F_4 generation. The process of ear-to-row selection is repeated for three or four more generations until the single-plant progenies are uniform and stable. Yield trials, and milling and flour tests, begin at the F_5 generation. Promising lines are entered for National List trials at F_9, and a new variety may be distributed twelve years after the initial crossing.

Selection with Outbreeders

Outbreeding species require a rather different approach for selection, because cross-pollination causes such species to share a common gene pool. When single plants are selected from an outbreeding population, segregation occurs and the progeny will differ from their parents in vigour, fertility and yield. It is therefore necessary to develop breeding strategies that lead to selection of a high frequency of favourable gene combinations without a reduction of vigour in important characters.

---

the basis of official trials; those that pass will be exposed to the decision of farmers on whether or not they will adopt them on the basis of economic merit. From Simmonds (1979) with permission.

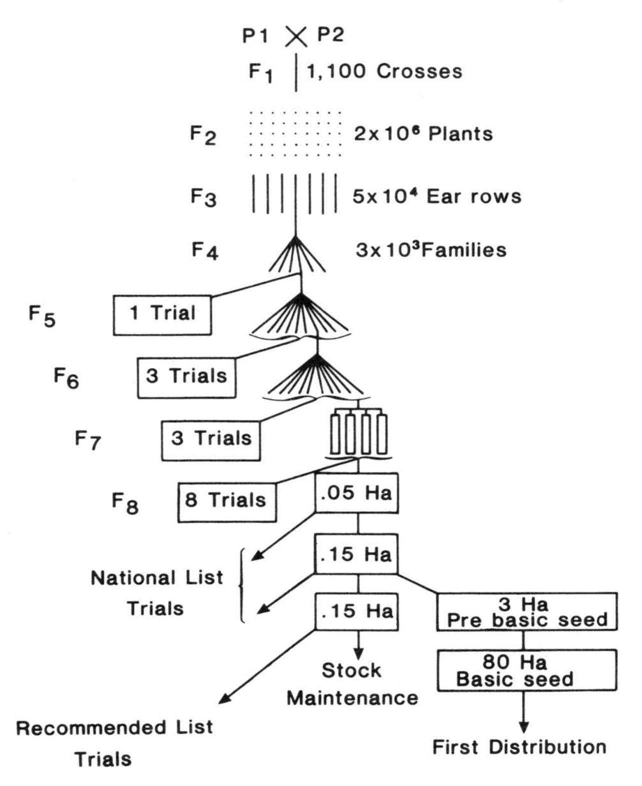

Fig. 1.2 Winter wheat selection and multiplication scheme, as used at the former Plant Breeding Institute, Cambridge. For description see text.

The result of cross-pollination is that each plant is heterozygous for many genes. Repeated inbreeding can result in loss of vigour or inbreeding depression. Two causes for this have been suggested. In the 'dominance hypothesis' it is assumed that deleterious genes are normally recessive, and their expression would be suppressed by the corresponding dominant alleles when in the heterozygous state. In the 'overdominance hypothesis', it is suggested that heterozygous combinations of alleles at a given locus, Aa, are superior to both the homozygous alleles AA and aa. Thus vigour should increase in proportion to the overall heterozygosity of a plant. The weight of evidence suggests that at most loci the dominance hypothesis applies and so there is an increasing tendency to turn outbreeders into facultative inbreeders to reduce the 'genetic load' of deleterious recessive genes and to expose the useful ones.

The strategies to be adopted for breeding outcrossing species are varied, and depend on whether the crop can be self-pollinated artificially, or is self-incompatible. If plants can be selfed, then a modified form of single-plant selection can be followed. Thus it may be possible to inbreed for a number of generations, from which the best lines are selected, and these are then intercrossed to reestablish sufficient heterozygosity. If a species is self-incompatible, then it is necessary to select a number of superior genotypes (about six), which possess the desired characteristics of that crop, and these are then intercrossed to produce a synthetic variety.

Some Strategies for Outbreeding Crops

The simplest approach for selection of outbreeders is 'mass selection'. Populations of the crop plant are grown, and the weakest plants removed. A variation of this, 'line breeding', involves the selection of superior plants following mass selection and the random intercrossing of seeds from these superior plants. The plants selected for intercrossing should not be too closely related to avoid inbreeding depression.

Alternatively, various forms of 'recurrent selection' can be applied. In its simplest form this involves selection of superior genotypes from a heterozygous population, propagating them by selfing, then intercrossing all the selfed progenies. Plants produced are then used for further cycles of selection and intercrossing. Probably the most effective form of recurrent selection in outbreeders is 'recurrent family selection'. A population of genotypes is crossed together, the progeny from each mother plant ('half-sib family') are evaluated and plants selected from the best family(ies) for the next generation of selection.

Backcross breeding (see below) may also be applied to outbreeding crops, except that a number of plants must be used as recurrent parents, rather than a single parent.

Backcross Breeding

In many cases a good existing variety could be improved by transferring a desirable character from another variety, line or related species. This can be accomplished by making a series of backcrosses of the donor line with the good existing variety (or recurrent parent), and selecting for the desired character at each generation. The character to be transferred in the successive backcrosses should be readily identifiable. After a series of backcrosses, the progeny will be heterozygous for the alleles it is desired to transfer, but homozygous for all the others. (However, this process is often complicated by co-transfer of undesirable properties that are closely linked to the desired character.) The last backcross generation is selfed, and with selection, some progeny will be homozygous for the genes to be transferred, but otherwise identical to the original good variety.

Quantitative Inheritance

As discussed earlier, although major genes that control identifiable characters are easier to manipulate, many characters present a continuum of phenotype, and are the product of the sum of expression of genes whose effects are too small to be

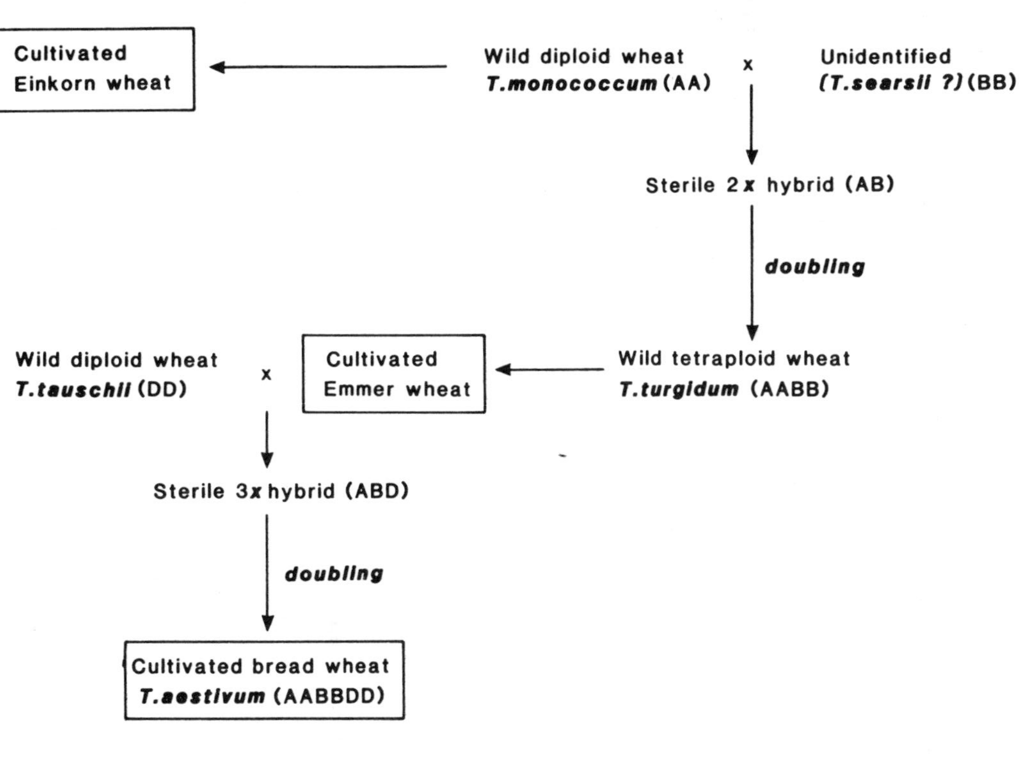

Fig. 1.3 Origins of the bread wheat genome. Three diploid wild grasses, each of which contributed a different genome (represented as A, B and D) have given rise to allohexaploid bread wheat ($2n=6x=42$).

identified individually. Such 'polygenes' do not easily exhibit measurable dominance, and polygenic characters can only be described in quantitative terms using 'biometrical analysis'. When considering polygenic systems, the aim of breeding is to bring together those genes that work in the same direction by selection for particular characters.

Polyploidy

Although diploid plants, with two sets of chromosomes in their nuclei, are considered as the normal form of eukaryotic organisms, plants with three or more complete sets of chromosomes are common, and these are referred to as 'polyploids'. These can be subdivided into 'autopolyploids', which arise through multiplication of the complete haploid or diploid chromosome set of the species, and 'allopolyploids', which contain chromosome sets from two or more different species. 'Euploid' plants contain complete sets of chromosomes, whereas 'aneuploids' contain more or fewer chromosomes than multiples of the basic number.

Chromosome doubling can be induced experimentally, and is usually achieved in plant breeding by applying the chemical colchicine, which disrupts spindle microtubules during cell division, so that daughter chromosome sets remain in the same cell. The main effect of chromosome doubling is to increase cell size, but the expression of specific genes is occasionally altered. Morphologically, polyploids tend to be larger than diploids, with thicker leaves, and they may respond differently to environmental conditions. Genetically, the plants may also differ. An autotetraploid with two allelic forms of a gene at a locus (A and a), can have five genotypes in the population (AAAA, AAAa, AAaa, Aaaa, aaaa). The population will therefore contain a much higher proportion of heterozygotes than homozygotes. Although polyploids with odd numbers of chromosome sets tend to be sterile, important odd-number polyploid crops do exist, such as bananas, some apple and pear varieties, and ornamentals. These are usually propagated asexually to overcome the fertility problems.

If two different diploid species are crossed, F_1 plants are usually sterile because the respective chromosomes are not homologous and cannot pair. If the chromosome sets are doubled, to give an allopolyploid, each chromosome can pair, and so the new hybrid species can behave as a diploid at meiosis, and such hybrids will be fertile. The best example of an allopolyploid crop plant is bread wheat (*Triticum aestivum*), which is an allohexaploid ($2n=6x=42$). The origin of bread wheat is given in Fig. 1.3.

Whereas autopolyploids possess genes similar to their diploid progenitors, allopolyploids combine the gene content of two different species, and hence their potential capacity for variation is greater.

Manipulation of Chromosomes

Aneuploids are of little direct value in plant breeding, but they can be used to locate genes on particular chromosomes, to establish linkage groups, and for

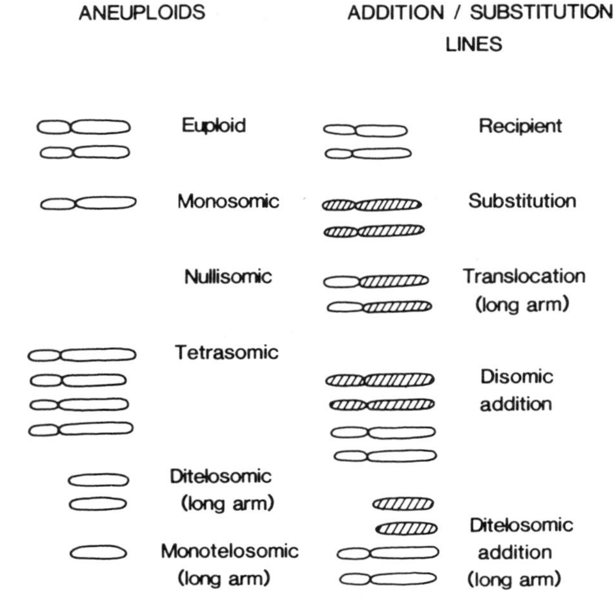

Fig. 1.4 Wheat stocks used in gene mapping. *On the left-hand side*, various aneuploid stocks are shown in addition to the euploid form with the normal pair of chromosomes. Thus 'monosomic' lines have lost one member of the chromosome pair, whereas in 'nullisomic' lines both members of the pair are missing. In a tetrasomic plant there is an extra pair of chromosomes, and in a ditelosomic or monotelosomic plant the short arms of the chromosomes have been lost, and two or one long arms respectively, are present. *On the right-hand side*, addition/substitution lines are shown. Either both chromosomes can be substituted for a different pair, or one arm of each pair may be translocated and substituted for a different arm. A new pair of chromosomes may be added ('disomic addition), or only individual arms of new chromosomes ('ditelosomic addition') (courtesy of Dr P. Shewry).

substitution of chromosomes (or parts of chromosomes) of one species for those of another.

In some cases, during meiosis, homologous pairs of chromosomes fail to separate, and pass to one pole of a cell instead of to opposite poles. If the result of fertilization of such gametes with a normal gamete is viable, then progeny have the chromosomal constitution of $2n+1$ ('trisomic') or $2n-1$ ('monosomic'). Aneuploids at the diploid level of this type are less likely to be viable than polyploids. Loss or gain of a chromosome leads to different phenotypes. By studying these types, the chromosomal locations of genes for useful traits, such as disease resistance or cereal storage proteins, can be determined. In some cases, notably for cereals such as wheat, the information can be used to substitute one or more pairs of chromosomes (Fig. 1.4), carrying agronomic traits (e.g. resistance to rust) from one genotype or species to another. This process is referred to as 'controlled introgression'.

Mutation Breeding

Gene mutation occurs naturally at a low frequency, in the order of 1 in 10^4 to 10^6 alleles for individual loci. Spontaneous mutation, followed by recombination, is the raw material of plant improvement, providing plant breeders with genetic variability from which to select. Because observable variation within a species in important agricultural traits may be limited, breeders have sought to increase the mutation rate of their material by a variety of physical and chemical methods. These include ionizing radiation (γ- and X-rays, α-particles and neutrons), non-ionizing radiation (UV-B light), and chemical mutagens such as ethyl methanesulphonate, diethyl sulphate, nitroso-compounds, nucleoside analogues and sodium azide.

There are various reasons why mutation breeding is attractive although its application has not been as successful as was originally hoped. These include: the possibility that completely new properties might be induced in an existing crop; that the expression of specific gene loci in an established variety might be induced, say, for disease resistance, so that a simply modified variety might be released rapidly after one or two generations of selfing or backcrossing; and that transfer of genetic material between species might be facilitated.

Over 460 commercial varieties have been produced following induced mutations, but barley has been the model plant of choice both for experimental mutagenesis and mutation breeding. Other crop plants treated in this way include peas, maize, wheat, rice, tropical crops and fruit trees. Since many mutations yield novelties, which are held at a premium in ornamental lines, mutation breeding has been extensively applied to ornamental flowering plants.

Mutation breeding has indeed led to the uncovering of new loci-controlling phenotypes, such as 'wax-less' variants of barley, and has led to altered grain storage proteins (e.g. to increase the proportion of the nutritionally limiting amino acid lysine in grains, as in 'Hiproly' barley; see Chapter 7). It can also cause loss of functions, and some of these, such as loss of self-incompatibility, can be useful in breeding. In some cases mutagenized plants have been directly utilized as varieties, and in others the mutations have been incorporated into varieties after crossing.

Male Sterility and Production of F_1 Hybrid Varieties

Male sterile plants, which occur naturally in a number of species, can be of great use to breeders to prevent selfing. In many cases male sterility, in which functional pollen is not produced, is governed by a single nuclear gene, and the condition can be maintained by crossing sterile males (SS) with heterozygous fertile plants (Ss), giving half male-sterile and half male-fertile progeny. Another type of male sterility that has been studied in some detail at the molecular level is 'cytoplasmic male sterility' (cms). This is usually associated with altered properties of mitochondria within the cytoplasm, so that the male sterile trait is inherited maternally. Cms plants are used as female parents in crosses, so that any seed formed on the cms parent will be the product of cross-pollination. For some crops (e.g. onion, sugar-beet, maize, sorghum) use of a cms parent is the only way to produce hybrid seed effectively and economically. Cms appears to result from nuclear–cytoplasmic interactions, and nuclear 'restorer' genes exist that can override the cms property and give fertility. For seed crops, restorer genes (R) are required. Thus crosses involve: female parent with cms cytoplasm and recessive (rr) nuclear restorer genes, and male parent with normal fertility and dominant restorer genes (RR). The F_1 hybrid will be heterozygous for the restorer gene (Rr) and so will be fertile and produce fruit or seed. The molecular basis of both cms and restorer genes is discussed in Chapter 8.

Improvements in Breeding Technology

There has been steady improvement in various aspects of breeding technology. These include: the introduction of alien variation by various techniques of sexual hybridization, combined with repeated backcrossing; an increase in the speed and precision of selection; and a decrease in the generation time required by achieving three or four generations per year in suitable environments. Breeding systems have also been modified (e.g. cross-pollinated changed to self-fertile), and improved definition of breeders objectives has speeded up selection processes.

Limitations to Established Breeding Practices

Although established breeding strategies have been remarkably successful in producing varieties with increased yields, resistance to diseases, pests and tolerance to stress, it has been estimated that only about 50% of the increased yield results from genetic factors. The remaining 50% has resulted from improved agronomic practices, including fertilisers, husbandry and crop protection sprays. Often genetic and husbandry factors are inseparable, such as the use of high levels of nitrogen fertilizers coupled with the growing of short-strawed wheat. The increase in yields, in such crops as wheat where gains have been most impressive, has on the whole been due to a change in the 'harvest index' (i.e. a change in partitioning of assimilates from the vegetative parts of the plants into harvestable organs), rather than an increase in biomass. Nevertheless, the yields of new varieties of forage crops, such as grasses, have been increased substantially (e.g. about 20%) over the past 30 years, even though the same strategy that has been applied to wheat cannot be adopted.

Given that established breeding practices have been successfully applied, what are the particular problems that make some aspects of breeding more difficult, and how can the developing biotechnologies be usefully applied? We can identify five particular breeding problems:

(1) *Genetic linkage.* One of the main problems that breeders must contend with is that of genetic linkage. The fact that certain genes show associated inheritance, because they are closely linked together in the same linkage groups (i.e. chromosome), leàds to a number of difficulties. Controlled introgression of genes from wild species is severely limited because it is difficult to detach unwanted from desirable genes. The same limitations occur when backcrossing, in response to recurrent selection, and even in mutation breeding, because desirable mutants are often linked to undesirable ones induced at the same time.

(2) *Polygenes from wild populations.* Because of the sheer number of genes that have to be re-ordered even if there is no linkage, breeders find it difficult to utilize polygenes from unadopted wild populations or different species.

(3) *Complex traits.* It is difficult to assess the many complex traits and the attendant need to identify their specific components, and to elucidate the genetics of those components so that they can be bred.

(4) *Identifying useful mutants.* There is a very low frequency of *useful* spontaneous mutants, and thus an attendant difficulty in identifying them, especially since they are usually recessive. (Hence the need to introduce genes from other species.)

(5) *Time-scales.* The very long time-scales of many breeding procedures, such as backcrossing, development of inbred lines, and genetic recombination by hybridization and selection, obviously slows production of improved varieties. This is particularly so for perennial crops such as fruit trees, which take many years to evaluate or complete a sexual generation.

Some of the factors that limit the progress of plant breeding may be overcome or achieved more rapidly by application of the new range of cell and molecular biological techniques that are currently being developed. In particular, new techniques of genetic engineering can make available to plant breeders genes from previously inaccessible organisms (e.g. viruses, bacteria, algae, animals and unrelated plants) by circumventing the normal processes of sexual reproduction both by transferring isolated genes directly and by culturing otherwise inviable hybrid embryos. Vegetative propagation can be extended to species for which this is not naturally available by exploiting *in vitro* culture systems. There are perhaps two most important longer-term effects of plant biotechnology: the generation of new genetic combinations and an increase in our understanding of genetic regulatory processes. The generation of genetic diversity will help overcome the increasing trend of current plant breeding which is to produce a standardized, uniform, proven product (demanded by industry and the consumer) but which is genetically restricted. The introduction of specific, valuable genes from distant or exotic species into major crops by novel techniques is a basic tenet of plant biotechnology in agriculture. An increase in our understanding of processes such as flowering and cell division and expansion through recombinant DNA technology, will also provide new opportunities for the application of existing breeding methods. It is the nature of these new techniques, the achievements which have already been made in their exploitation, and prospects for their future use, which is the subject of the following chapters.

General Reading

Austin, R.B. (1986). *Molecular Biology and Crop Improvement.* Cambridge, Cambridge University Press.

Lawrence, W.J.C. (1968). *Plant Breeding. Studies in Biology*, no. 12. London, Arnold.

Mayo, O. (1987). *The Theory of Plant Breeding*. Oxford, Oxford University Press.

Rees, H., Riley, R., Breese, E.L. and Law, C.N. (1981). 'The manipulation of genetic systems in plant breeding', *Phil. Trans. R. Soc. Lond.* **B292**, pp. 399–609.

Specific Reading

Simmonds, N.W. (1979). *Principles of Crop Improvement.* London, Longman.

Chapter 2

The Biology of Cultured Plant Cells

Introduction

The emergence of a plant biotechnology, and optimism for its contribution to the production and introduction of novel cultivars, or even of bypassing the plant altogether such as for the synthesis of natural products *in vitro*, has been the consequence of advances in techniques of plant cell and tissue culture and plant molecular biology. Before considering how these can be applied to biotechnological problems, we should consider briefly the biology of the systems involved. The growth and development of plants is controlled by interactions between the genome and the environment, and the term 'environment' can be taken to include the intra- and inter-cellular milieu, and also the ambient conditions of light, temperature, nutrient and water supply, and so on. Within the organism, intercellular communication is essential for normal plant function, and is effected by both short-distance and long-distance mechanisms. Cells can, at one level, be considered as discrete structures, each bounded by a cell wall and each type having a specialized function; but the presence of plasmodesmatal connections means that the cytoplasms of all cells of the plant, apart from the mature pollen grains and the embryo, are in contact. Other more specialized forms of short-distance communication include the cytomictic channels of meiocytes and recognition systems, such as between pollen grains and stigma, pollen tube and egg cell, and root and root nodule. Long-distance communication involves the transport of growth regulators, ions, primary and secondary metabolites and water via the xylem and phloem. The roles of such communication systems are of obvious importance in plant differentiation and development, but how autonomous are the constituent organs, tissues or cells? A number of species, in the wild, are capable of regeneration from small pieces of severed tissues—dandelions proliferate from isolated roots, and *Begonia* plantlets develop directly from leaf tissue—and this phenomenon of vegetative propagation demonstrates clearly the plasticity of plant development. It was an interest in defining the functional relationship of one tissue with another, and of investigating the potential for development of cells in isolation from the intact plant, that gave rise, at the turn of the nineteenth century, to the science of plant tissue culture.

In 1902, Gottlieb Haberlandt elegantly related his vision of the future of cell biology:

> To my knowledge, no systematically organized attempts to culture isolated vegetative cells from higher plants in simple nutrient solutions have been made. Yet the results of such culture experiments should give some interesting insight into the properties and potentialities which the cell, as an elementary organism, possesses. Moreover it would provide information about the inter-relationships and complementary influences to which cells within the multicellular whole organism are exposed.

Haberlandt himself worked with single cells isolated from the palisade layer of leaf tissue, with parenchyma, and with the epidermis and epidermal hairs of a number of species, but he was never able to induce cell division *in vitro*. Little further progress was made until 1934 when White cultured tomato roots on a simple medium of inorganic salts, sucrose and, importantly, yeast extract. In the same year Gautheret found that cambial tissue of *Salix capraea* and *Populus alba* could proliferate for several months after aseptic isolation, but growth was limited. Over the next five years or so, however, the recognition of the importance of B vitamins and the auxin indole-3-acetic acid (IAA) allowed significant advances to be made, and in 1939 Gautheret reported the propagation of the first plant tissue culture of unlimited growth, a strain of carrot isolated two years previously.

Since these historic studies, the nutritional requirements for the proliferation, *in vitro*, of plant cells of a large number of species have now been established, and much work has been carried out to characterize the biochemistry, physiology and cytology of cultured cells. Here, we briefly discuss some of the features of *in vitro* cultures, beginning with the first cell divisions of the explant tissue. When a tissue is physically damaged, viable cells at the wound site will normally start to divide as a 'wound response'; and when the wound is sealed up, the cell divisions cease. A similar situation is observed when a graft union is made between compatible species—after an initial series of cell divisions at the two adjacent cut surfaces, cell differentiation occurs, resulting in the establishment of vascular continuity between the stock and scion. If, however, wounded tissues are isolated from the plant, and instead maintained on an artificial nutrient medium, the natural regulatory mechanisms which limit the extent of cell divisions can be over-ridden, and callus proliferation will continue as long as adequate nutrients and growth regulators are supplied (Fig. 2.1).

The production of a callus from an explant tissue is accompanied by a series of

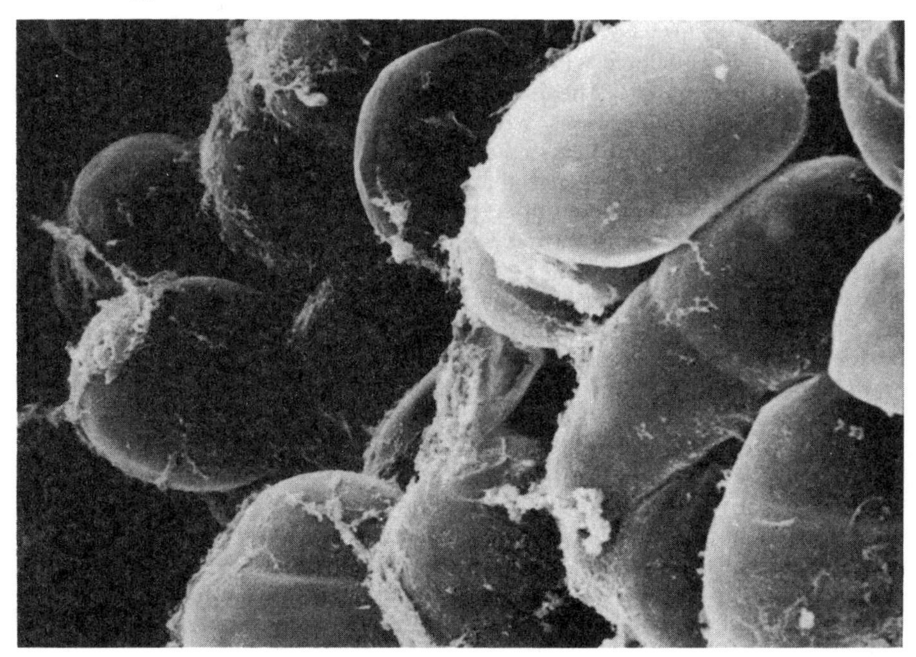

Fig. 2.1 Undifferentiated callus cells of carrot, as seen under the scanning electron microscope.

dramatic changes in the appearance and metabolism of the cells (Aitchison *et al.*, 1978; Yeoman and Street, 1978), although the precise nature of the response to culture conditions is influenced by both the composition of the nutrient medium and the physiological characteristics of the explant tissue. A large amount of the early work on the wound response was carried out on tuber slices of the Jerusalem artichoke, in which the first divisions at the cut surface were largely synchronous. Division is initially restricted to the peripheral layers of the explant, with the formation of a wound cambium, the cells of which divide but do not expand appreciably. Changes occur at the ultra-structural level which reflect an alteration in the metabolic activity of the cells. An increase in respiration is associated with increased numbers of mitochondria and increased activities of glycolytic enzymes such as hexokinase, glucose-6-phosphate dehydrogenase and malic dehydrogenase. Storage products such as starch grains disappear, and the synthesis of RNA and protein is reflected in the appearance of increased numbers of ribosomes and polyribosomes. By four to six weeks of culture, the explanted tissue may have produced its own weight of callus tissue, which, if excised and placed on fresh nutrient medium, will continue to grow.

The production of a callus can be considered to be the de-differentiation of an organized tissue, as determined by the observed changes in morphology and metabolic activity. Under appropriate nutritional and hormonal conditions, callus morphology may be described as either friable (in which the cells are only relatively loosely associated with one another) or as compact (when the cells are more densely aggregated). The different types of callus may be characterized by different cell-wall compositions, with compact callus cells possessing a higher proportion of pectins and hemicelluloses in their walls than the friable type; however, the two types are generally not specific to a particular species, being interchangeable and dependent upon the medium composition. Friable callus is usually readily dispersed to form a suspension culture when transferred to an agitated liquid medium.

Associated with the physical disorganization of cultured cells is the disorganization of the metabolism characteristic of the explant tissue. This would appear to be due, in part, to the breakdown of intercellular communication (both physical and chemical) which characterizes the intact plant, and has been a source of frustration for those research workers interested in using cultured plant cells to synthesize specific natural products: in the majority of cases, callus or suspension culture cells derived from an organ which synthesizes a particular secondary metabolite fail to synthesize that compound to a similar level; this subject is further explored on p.70. Furthermore, although callus and suspension cultures may possess chloroplasts and chlorophyll, and indeed may in some cases be demonstrably capable of fixing gaseous carbon dioxide, fully autotrophic lines are uncommon, and are obtained only after rigorous selection procedures in CO_2-enriched atmospheres (Hüsemann, 1985). Photosynthetic capacity is suppressed by the presence of sucrose in the medium at substrate concentrations, in part, at least, due to a suppression of chloroplast lamellar development, and of the activities of enzymes associated with chlorophyll synthesis, and also to a reduction in the numbers of mature chloroplasts per cell.

GROWTH CHARACTERISTICS

The growth of relatively undifferentiated callus and suspension cultures can be quantified in a number of ways, but most commonly as an increase in fresh weight, dry weight or cell number over a time course. A generalized growth pattern takes the form of a sigmoidal curve (Fig. 2.2), in which can be recognized a lag phase, in which there is little or no cell division, a phase in which the proportion of dividing cells is at a maximum (the 'exponential' phase), followed by the linear and stationary phases, in which the proportion of dividing cells gradually declines and vacuolation proceeds. Each stage can be characterized by distinct structural and biochemical features, but the whole growth cycle is in reality a continuum of physiological changes, and at any particular time point there will be found, not a homogeneous population of identical cell types, but a spectrum of cells of different morphology, biochemical behaviour and genetic complement. The consequences of the tissue culture process in cytological and biochemical terms (somaclonal variation), and some of the benign spin-offs of this phenomenon, will be described and discussed in Chapter 5.

Before considering some of the biological features of cultured plant cells, let us briefly summarize the types of methods used for initiating cultures. For more

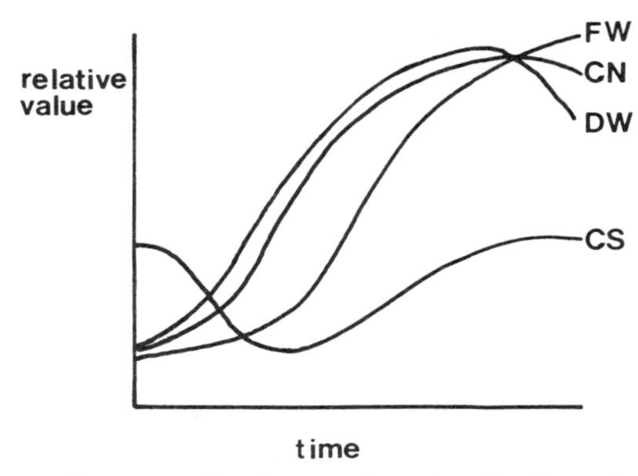

Fig. 2.2 Features of a generalized growth cycle of cultured plant cells. FW = fresh weight; CN = cell number; DW = dry weight; CS = mean cell size.

detailed literature concerning the methodologies for specific species, the reader is directed to the various excellent handbooks listed on p.33.

Callus Culture Initiation

As we have seen, the damage of plant tissue usually results in the initiation of cell division at, or adjacent to, the cut surface, and if the tissue is explanted and supplied with a mixture of salts, vitamins, growth regulators (usually an auxin and a cytokinin) and a carbon source (usually sucrose), cell division will continue and callus will be formed. A range of tissues (in fact, probably any tissue) can be used successfully as an explant material, but the tissue of origin of the explant may have a profound effect on the subsequent developmental behaviour of the callus culture —this important phenomenon is discussed below. In general, young or developmentally immature tissues such as hypocotyls, young leaves, immature embryos or immature inflorescences prove to be a suitable starting material. A procedure for the initiation of a callus culture from tobacco stem tissue is summarized in Fig. 2.3. A section of stem, approximately 3 cm in length, is cut from the youngest part of a healthy plant, and the cut ends are sealed with molten paraffin wax. This sealing prevents internal damage to the stem during the subsequent sterilization process, in which the surface is cleaned of oils and 'partially' sterilized by immersion in 70% ethanol for 20–30 sec, followed by complete sterilization in a halogen-containing solution such as sodium hypochlorite (approximately 2% available chlorine). Other sterilants available include calcium hypochlorite, mercuric chloride and bromine water. The presence of one or two drops of a surfactant such as Tween-80 in the sterilant solution is recommended to improve the contact between tissue and sterilant. After rinsing in five or six changes of sterile distilled water, the stem is cut into discs approximately 2–3-mm thick (the wax-coated ends being discarded) and each disc is placed, cut-face upwards, on an agar-solidified medium such as Murashige and Skoog medium containing $30\,g\,l^{-1}$ sucrose,

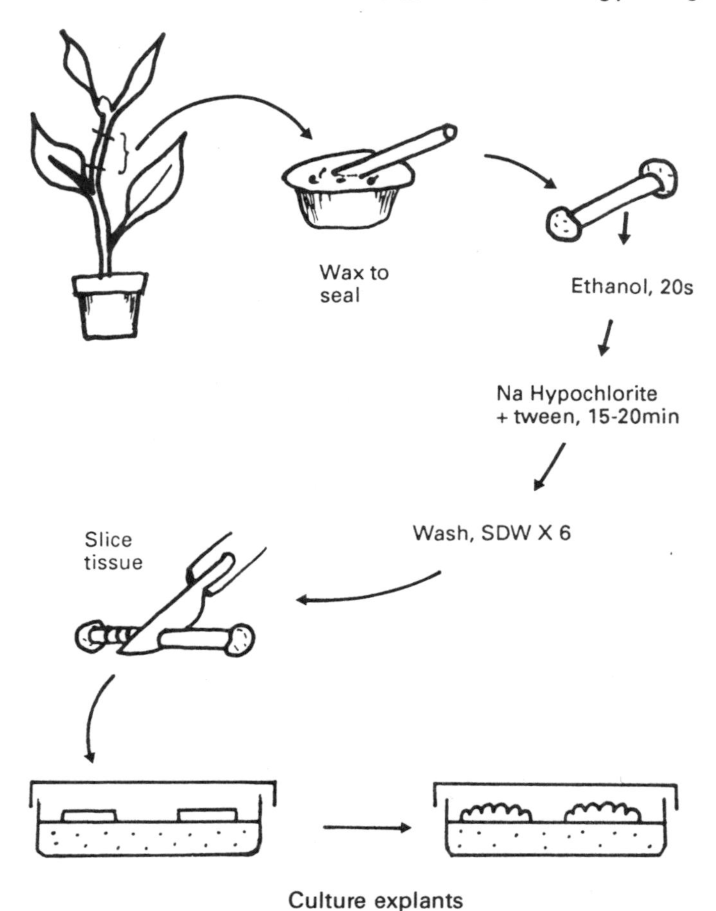

Fig. 2.3 A method for the initiation of stem callus cultures. (See text for details.)

$1\,mg\,l^{-1}$ 2,4-dichlorophenoxyacetic acid (a synthetic auxin) and $0.2\,mg\,l^{-1}$ kinetin (6-furfurylamino purine, a synthetic cytokinin). The tissue is usually maintained under fluorescent lighting ($10–100\,\mu mol\,m^{-2}\,s^{-1}$) at 24–28°C, and the first evidence of callus formation can be recognized as a crystalline appearance of the cut surface within 1–3 weeks. After 4–8 weeks, enough callus will have developed to be transferred to fresh medium, and the culture is established.

SUSPENSION CULTURES

Suspension cell cultures are initiated by transferring preferably friable callus to liquid nutrient medium which is agitated, usually on an orbital shaker at between 90 and 150 rpm, with an orbital diameter of 1–2 cm. Agitation serves both to aerate the culture and to disperse the cells which, either as a result of cell division

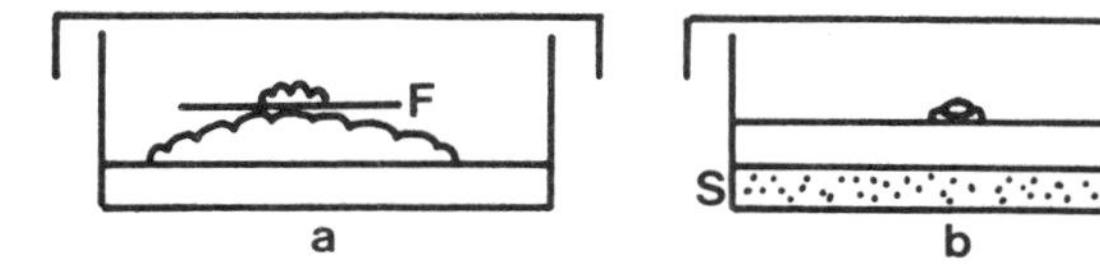

Fig. 2.4 Two methods of nurse culture: (a) Single cells are cultured on the surface of a filter paper (F), itself in contact with a nurse callus. (b) Single cells are cultured on a layer of agar medium which has been poured on top of agar medium in which are embedded nurse suspension culture cells (S).

at the callus surface or of mechanical effects, break away from the inoculum tissue. The division rates of suspension culture cells at the exponential phase are typically higher than those of callus cells, but doubling times are nevertheless slow in comparison with those of microbial cells, and are usually within the range of 24 to 72 hours. It is a common observation that if relatively small numbers of cells are transferred to a new medium (either solid or liquid), they may fail to divide; whereas a larger quantity of tissue transferred from the same culture may proliferate rapidly on the same medium. This observation has led to the concept of a 'critical initial cell density', which can be defined as the smallest inoculum, per volume of medium, from which a new culture can be reproducibly grown. Factors which influence the minimum size of inoculum include (a) the culture's physiological characteristics, (b) the length of time and conditions under which the culture was previously maintained, and (c) the composition of the fresh medium. This last point is of considerable interest, for it has been frequently reported that single cells or protoplasts can be induced to divide (a) if they are cultured in close proximity to, for example, a large piece of callus, albeit separated from the cell by a porous membrane or filter paper – in this situation the callus is referred to as a 'nurse tissue' (Fig. 2.4), or (b) if the medium, which has previously been used to maintain a culture, is used to supplement the new medium supporting the single cell. This effect is described as 'conditioning', and a conditioning medium derived from an exponential phase culture is usually the most efficacious. These observations indicate that aggregates of cells ('feeder cells') release chemical substances into the medium which promote cell division and growth; animal cell cultures exhibit the same phenomenon. The chemical basis of conditioning is uncertain, and cannot in most cases be mimicked simply by adding to the medium extra nutrients or growth regulators. Nevertheless, both these approaches have been used successfully to culture and regenerate a number of previously recalcitrant species, notably monocotyledonous species such as rice and maize.

Phases of the Growth Cycle

Once established, the maintenance of callus and suspension cell cultures proceeds as a series of growth cycles. Each cycle is characterized by its 'sigmoidal' nature, comprising a *lag phase*, a *cell-division phase* to a *stationary phase*. These will now be discussed in turn.

THE LAG PHASE

When cells are subcultured in a fresh nutrient medium, there is initiated a series of metabolic processes which prepare the cells for mitosis. Increases occur in the levels of reducing power, ATP and energy charge, as carbohydrate, commonly supplied as sucrose, is oxidized by glycolysis, the pentose phosphate pathway and the tricarboxylic acid (TCA) cycle. There are also increases in the rates of synthesis of proteins and nucleic acids, and the steady-state levels of specific mRNAs have been shown to be strongly influenced by the composition, particularly by the auxin type and concentration, of the medium. Other transient increases in the activities of specific enzymes, notably of phenylalanine ammonia-lyase, appear to be induced simply as the result of a cell-dilution effect (Schröder *et al.*, 1977).

THE CELL-DIVISION PHASE

Once cell division has been induced, it will proceed until one or more of the nutrients becomes limiting. The increase in cell number usually precedes the increase in fresh weight, due to the time delay for cell expansion. This results in a sharp decrease in the mean cell size during the exponential phase (Fig. 2.2), when the division rate is the greatest: Yeoman and co-workers (1965) have shown that, over the first 7 days of culture, Jerusalem artichoke explants can exhibit a 10-fold increase in cell number when cultured at 25°C. Such a burst of activity is associated with an increase in respiratory activity (measured as O_2 consumption) and steady-state RNA levels during the first 3 days of culture—but there is a decline of RNA levels on a per-cell basis. Morphologically, cells from the exponential phase appear undifferentiated as division outpaces expansion, with small vacuoles (cf. the meristematic cells of stem apices) and typically accumulate only very low levels of secondary metabolites (see below). During the linear phase of the growth cycle, the mean cell size changes little, and rates of protein and nucleic acid synthesis remain relatively high.

STATIONARY PHASE

The stationary phase is characterized by a decline in the rates of cell division, respiratory activity, RNA and protein synthesis, and an increase in the proportion of vacuolated cells. The cessation of division and the onset of expansion is in many examples associated with the synthesis and accumulation of specific secondary metabolites, including some alkaloids, anthocyanins and other phenolic derivatives, and anthraquinones. Some of the regulatory aspects of this phenomenon are considered later in relation to the biotechnological exploitation of cell cultures for the synthesis of secondary products. In cultures of some species, notably those of the Solanaceae, this biochemical differentiation may be accompanied by limited structural differentiation, especially if a fresh medium, conducive to cell proliferation, is not supplied. For example, if callus cultures of *Atropa belladonna* or *Hyoscyamus niger* are not subcultured at week four of the growth cycle, then, over the subsequent 4 weeks there may develop roots, shoots or embryo-like structures.

Similar embryoids may also develop in late stationary-phase suspension cultures, and the tropane alkaloids, such as atropine and more commonly hyoscyamine and scopolamine, may be synthesized and accumulated in these differentiated structures or even in simple cell aggregates (Lindsey and Yeoman, 1983). So, as nutrients become exhausted, limited reorganization and redifferentiation of the cell takes place and is reflected by secondary metabolic activity more similar to that of the intact plant. This phenomenon serves to illustrate the dependence of metabolic integrity on structural integrity.

Manipulation of Growth and Differentiation *In Vitro*

The aim of the above discussion has been to reveal the dynamic nature of a plant cell culture. In response to the nutritional and hormonal conditions, a series of metabolic and physiological processes is activated. As the cells take up and release metabolites they directly influence the composition of the medium, and adapt, in terms of the rate of cell division and of biosynthetic behaviour, to the continuously changing environment. It is possible, of course, to make specific, experimental modifications to the medium composition, and in doing so, direct the developmental fate of the cells. This is by virtue of the plasticity of plant cell development and the observed genetic totipotency of fully differentiated cells (although in practice it may be extremely difficult to demonstrate this phenomenon for a given single cell). Nevertheless, the manipulation of differentiation is probably the most important single approach in plant biotechnology as a means both to regenerate whole plants from single cells and to induce the synthesis of secondary metabolites. We will briefly consider the biological basis, and some of the specific techniques, of the procedures involved.

In a series of classic experiments, Skoog and Miller (1957) demonstrated an apparently simple relationship between the auxin–cytokinin balance of the nutrient medium, and the pattern of redifferentiation of unorganized tobacco pith callus. It was observed that, if the cytokinin (kinetin) concentration was high relative to that of auxin (IAA), then shoot development was induced; whereas if the auxin concentration was relativly high, then roots would develop. At intermediate concentrations, the pith tissue developed as an unorganized callus (Fig. 2.5). Although the precise requirement for auxins and cytokinins may differ between species, this general approach, namely the manipulation of exogenous auxin and cytokinin levels, has formed the basis of regeneration techniques for a large range of species. The experiments of Skoog and Miller also illustrate the importance of the role of the growth regulators, over and above that of other components of the medium, in eliciting particular developmental pathways. The mechanism or mechanisms of action of growth regulators is far from clear—it may be, in general terms, that these substances have a primary role in determining the developmental fate of the cells upon which they act, or that they act as triggers to activate a developmental pathway to which the cells are already committed (Meins, 1986). What is known, however, is that the activity of a particular combination of growth regulators and, to a lesser extent, other medium

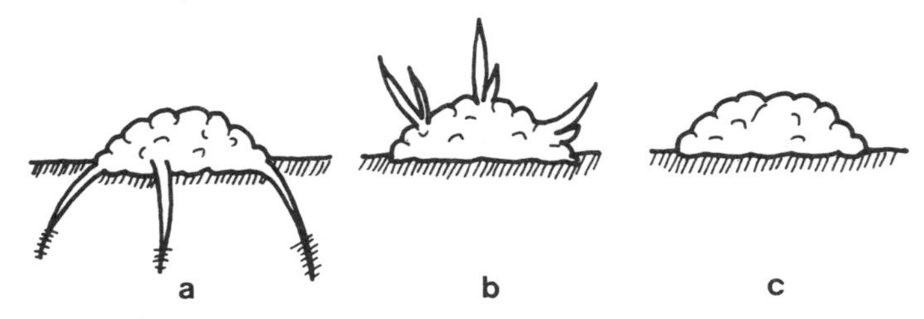

Fig. 2.5 Organogenesis in dicots induced by different ratios of auxin to cytokinin. (a) High auxin: cytokinin leads to root formation. (b) Low auxin: cytokinin leads to shoot formation. (c) Intermediate ratios lead to callus proliferation.

components, depends on the physiological 'history' of the cultured cells, including the tissue of origin of the explant, and the length of time the cells have been maintained *in vitro*. Morphogenetic and biosynthetic capacity is often found to decline over successive growth cycles, accompanied by (but not *necessarily* the result of) increased genetic variability, a subject which we will return to later on (see p.78). Let us examine some of the effects of the explant on the behaviour of cells cultured from it.

The Role of the Explant

As indicated earlier, the best growth of callus cultures is, in general, obtained if the explanted tissue is derived from a part of the plant which itself is either rapidly growing or is at an early stage of development, such as, for tobacco, the youngest part of the shoot. If explants are taken from progressively older regions there may be observed to be a gradient of declining ability to produce callus *in vitro*. These differences, which may be stably maintained, but are believed not to be the result of any genetic differences within the plant, have been described as 'epigenetic'. A classic example of such an epigenetic phenomenon is the habituation of tobacco cells. Habituated cells possess the ability to be cultured in the absence of auxins (auxin habituation) or cytokinins (cytokinin habituation). The latter type, which has been studied in some detail, is due to an alteration in the regulation of cytokinin synthesis, and can arise either spontaneously in cultures, or can be induced by the application of high temperatures (35°C) or low levels of kinetin (Meins *et al.*, 1980). Due to its high frequency and reversibility, such as by low-temperature treatment, cytokinin habituation is not considered to be the product of a genetic mutation, but may nevertheless be stable under a particular set of environmental conditions. Interestingly, the frequency at which tobacco callus becomes (or can be induced to become) habituated depends on the part of the plant from which the explant material was derived—the greatest rate of habituation is in cultured cells of explant tissue obtained from close to the shoot tip, and there is a gradient of declining habituation tendency as explants derived

Fig. 2.6 Mature and juvenile stages of ivy. The juvenile form (a) is characterized by lobed leaves, a climbing growth habit, an absence of flowering and anthocyanin pigmentation. The mature form (b) is characterized by ovate leaves, an orthotropic growth habit, the presence of flowers, and an absence of anthocyanins.

from older tissues are used. A second example of the epigenetic influence of the explant on callus behaviour has been described in cultures of ivy (*Hedera helix*). Ivy plants may have two types of leaf on different parts of the individual, each associated with the 'juvenile' and 'adult' phases of the plant respectively, but being genetically identical (Fig. 2.6). Callus cultures initiated from either part of the same plant exhibit stable differences in morphology and growth: 'juvenile' callus grows more quickly, and is composed of larger cells, than 'adult' callus, and characteristically produces fewer roots on a high salt medium.

At a more biochemical level, it has been observed that the isoenzyme pattern, and the profile of activities of specific enzymes, may be different in callus cultures derived from different tissues of the same plant. It would seem, therefore, that cells from different explants do not 'de-differentiate' in precisely the same way. In the words of Meins (1986)

> There is growing evidence that cultured tissues are not a 'blank slate'. Rather, they sometimes remember where they came from in the plant and this memory influences morphogenesis and genetic stability in culture.

Unfortunately, the 'neurological' basis of this developmental memory is as yet undefined; however, a clarification of the mechanisms involved would be of great value to those interested in unravelling the problems of regenerating 'difficult' species, such as some monocots and woody plants.

Patterns of Structural Organization

Reorganization of callus or suspension cells commonly proceeds by either the formation of organs (shoots or roots) or of bipolar embryo-like structures (embryoids, somatic embryos). We have already seen how, for some species at

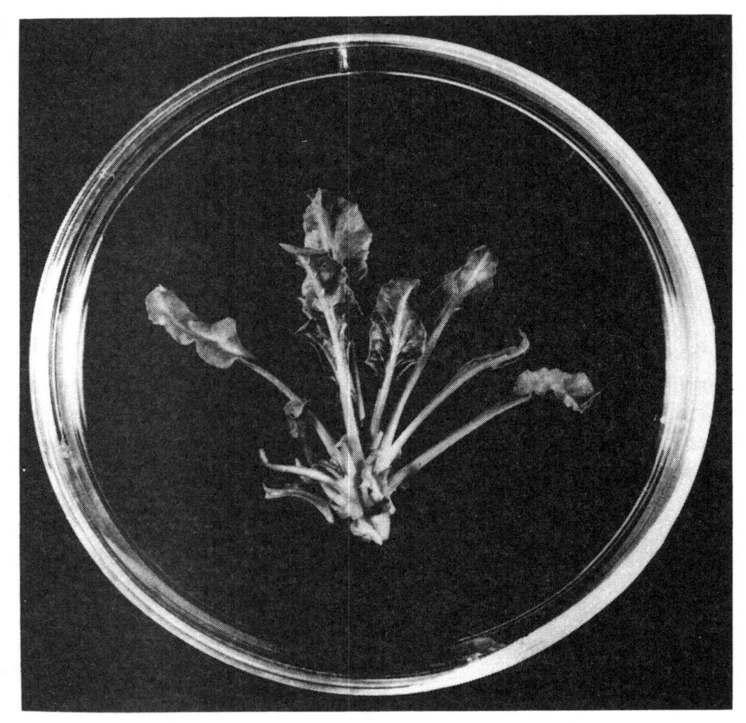

Fig. 2.7 Shoot culture of sugar-beet (photograph kindly provided by C. Eady).

least, organogenesis can be manipulated by experimentally altering auxin: cytokinin ratios. It is also possible to establish an organ culture with the minimum of unorganized growth by isolating explants from meristematic tissues, such as the shoot apex or axillary buds – here the culture is characterized by the proliferation of recognizable morphological structures, usually either shoots *or* roots (Fig. 2.7). By an appropriate manipulation of the medium composition, it may be possible to induce further morphological development, such as the formation of roots on cultured shoots to create a functional intact plant, and this area is discussed in more detail in relation to micropropagation in Chapter 4. It is still a question of debate whether organogenesis originates from a single cell or from a group of cells. Inoculation of explants with *Agrobacterium*, in transformation studies, usually results in the subsequent regeneration of genetically-uniform plants rather than chimaeras, and this has been considered to be evidence that organogensis is from a single transformed cell. However, it has not been demonstrated that the simultaneous transformation of a group of morphogenetic cells is not possible. Cytological studies of calluses and explants have revealed nodules of meristematic activity, suggesting that an area of tissue dividing in an organized fashion contributes to a developing organ.

Embryogenic cultures can be produced from explanted immature embryos, inflorescences or leaves, an observation which has revolutionized regeneration techniques for cereals and other grasses (Vasil, 1987). The application of the approach of Skoog and Miller, namely to induce organogenesis in an undifferentiated callus by manipulating the ratio of auxin to cytokinin in the medium, was largely unsuccessful for these species. However, somatic embryogenesis was observed in callus of a particular morphology, termed 'type-I' callus, characterized by compact, slow-growing, white to pale yellow tissue of small, densely-cytoplasmic cells, which was only produced from particular explants under specific hormonal conditions. Non-embryogenic callus is soft and friable, composed of vacuolated thick-walled cells. In some species, such as maize, a rapidly-dividing and friable, but nevertheless embryogenic, callus has been described, and this has been designated 'type-II' callus. Cereal species require high levels of the synthetic auxin 2,4-D for the induction of embryogenic callus, which if embryos are used as explant material, is derived from peripheral cells of the scutellum; if immature inflorescences are used, callus is derived from the floral meristem or ground tissue close to the vascular bundles, and if leaves are used, from cells of the lower epidermis and mesophyll close to the vascular tissue. In reality, embryogenic cultures are a mixture of embryogenic and non-embryogenic cells, and the embryogenic type is maintained by selective sub-culturing on to a high 2,4-D-containing medium.

If embryogenic callus is transferred to a medium with reduced 2,4-D, the formation of somatic embryos will occur. Development is by divisions of a single cell or of a pro-embryonic mass, resulting in a bipolar structure which in turn establishes two apical meristems. This process is similar in cereals and in dicotyledonous species, such as carrot or celery, and bears strong similarities to zygotic embryogenesis, discussed in Chapter 8. It is possible to increase the frequency of embryogenesis in a suspension culture by enriching for embryogenic cell aggregates. This can be achieved by sieving the culture through nylon or stainless-steel mesh or through columns of glass beads, or by density-gradient centifugation to size fractionate the cell population (Fujimura and Komamine, 1984). On transfer to, for example, an auxin-free medium some degree of synchrony in embryogenesis can be obtained from embryogenic aggregates.

More Specialized Techniques: Immobilized Cells and Protoplasts

Before concluding this chapter on the biology of cultured plant cells we will introduce two, more specialized, techniques which are of biotechnological significance: namely techniques of cell immobilization and of protoplast production.

IMMOBILIZATION

Immobilized plant cells are characteristically cultured as multicellular aggregates, supported by some form of biologically inert substratum (Lindsey and

Yeoman, 1986). The substratum is either physically fixed in relation to the liquid medium, which can wash over the cells, or the immobilized particles can be suspended in the medium. The most common intended use of immobilized plant cell systems is in the large-scale production of secondary metabolites, either by replacing immobilized enzymes or microbes as catalysts in one- or two-step biotransformation reactions, or in more complex, multienzyme syntheses. Protoplasts (see below) are also commonly cultured in the immobilized state to improve their division and growth characteristics and as an aid in the selection of transformed or mutant protoplasts, where frequent changes in the medium may be required and are facilitated if the protoplasts/cells are immobile.

The methods available for the immobilization of plant cells are in essence derived from those originally developed for microbial and animal cells. Two general techniques are most commonly used, namely embedding and entrapment (Table 2.1). In the first, cells or protoplasts are embedded in a polymeric gel or in a combination of gels, such as calcium alginate, agar or agarose. The viability of the cell preparation depends initially upon the type of gell matrix used—alginate gels are considered to be particularly mild, whereas others such as polyacrylamide are often toxic and are not widely used for plant cells. Entrapment techniques involve either the immobilization of cells in pre-formed foams or meshes, or in hollow-fibre membrane systems. In the hollow-fibre systems, tubular fibres of, for example, cellulose acetate or silicone polycarbonate are arranged in cartridges in parallel bundles. Cells are trapped in the spaces between the fibre membranes which are permeable to the liquid medium passing along the fibre cavity. An increasingly used entrapment matrix is, however, reticulate polyurethane foam (Fig. 2.8), in which freely-suspended cells readily become trapped and grow as aggregates while retaining their viability. The role of immobilized cell techniques in plant biotechnology is discussed in Chapter 4.

PROTOPLASTS

Protoplasts are produced by the enzymatic removal of the cell wall, using mixtures of commercially prepared fungal cellulases, pectinases and hemicellulases in a solution of high osmotic potential (Table 2.2). It is possible to isolate protoplasts from both intact plant tissue (commonly leaves) and also from suspension culture cells, and a generalized protocol for leaf protoplast isolation is illustrated in Fig. 2.9; many variations on this methodology are possible (see Vasil, 1984), and the precise conditions for each species and source tissue must be determined empirically. Essentially, non-sterile plant material is surface sterilized and rinsed, and the cells are plasmolysed in a solution of mannitol, sorbitol or sucrose of high osmotic potential (700–800 mOsm), to free the cytoplasm from its close association with the cell wall and to break plasmodesmatal connections between cells. Pectinase treatment separates individual cells by dissolving the middle lamella and pectins within the wall, and allows access of cellulases and hemicellulases. Released protoplasts may be separated from undigested cell aggregates by sieving through stainless-steel or nylon mesh (38–100 μm pore size), and collected after washing by either flotation on 21% sucrose or by centrifugation on a sucrose

Table 2.1 Some techniques for plant cell immobilization

Embedding	*Entrapment*
Calcium, barium alginate	Reticulate polyurethane
Agarose	Hollow fibres
Agar	Polyphenyleneoxide
Carrageenan	Fibrous polypropylene
Gelatin	Lectins
Polyacrylamide	
Alginate and gelatin	
Agarose and gelatin	
Polyacrylamide and alginate	
Alginate and nylon	
Hypol-3000	

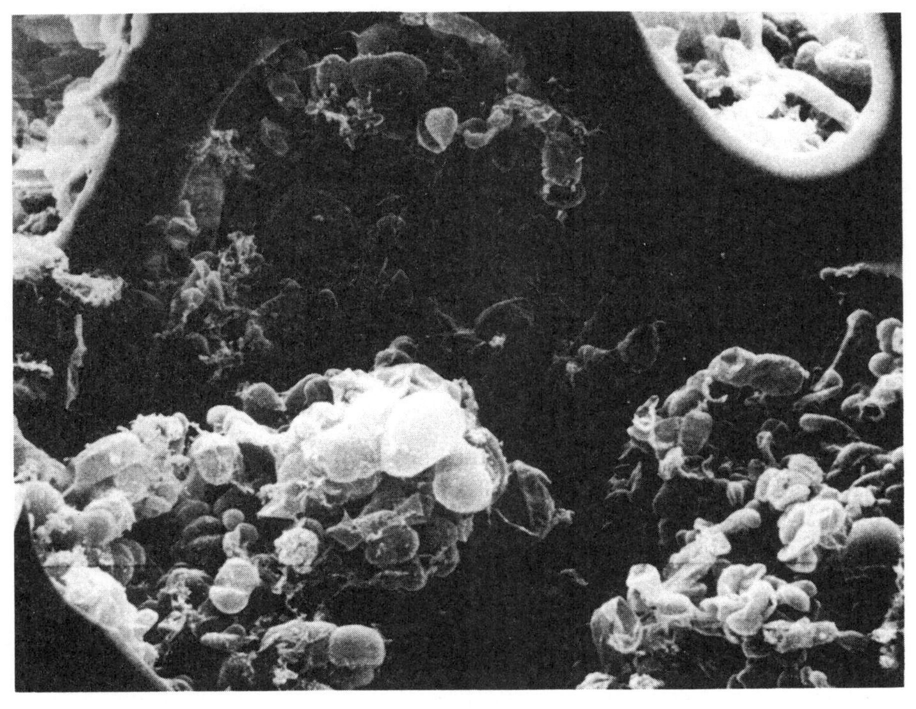

Fig. 2.8 Cells of *Capsicum frutescens* immobilized in reticulate polyurethane.

Table 2.2 Cell wall-digesting enzymes

Hemicellulases	*Pectinases*	*Cellulases*
Hemicellulase	Pectolyase Y-23	Meicelase
Rhozyme	Macerozyme R-10	Cellulase R10
		Cellulase RS
		Driselase

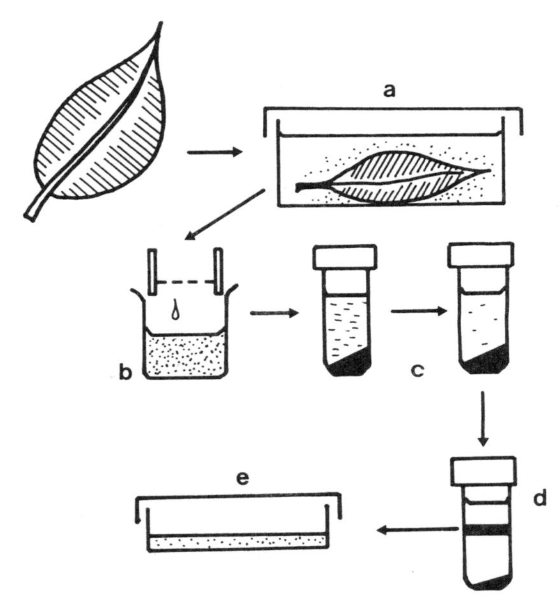

Fig. 2.9 Procedure for leaf protoplast isolation. (a) Sliced leaves are incubated in an enzyme cocktail to release protoplasts. (b) Undigested material is removed by sieving and the protoplasts are (c) washed, (d) purified on a sucrose cushion and (e) cultured in medium solidified by agarose.

cushion—cell debris is pelleted, while the buoyant protoplasts float at the sucrose surface. It is possible to culture protoplasts (Fig. 2.10), which normally develop a cell wall in two to three days, to produce callus cultures and even to regenerate into whole plants, but this entire cycle of reorganization has been achieved only in relatively few species.

A variety of factors influence the yield and subsequent 'performance' of protoplasts (their ability to divide, to form microcallus colonies and to regenerate), and a list of some of these is presented in Table 2.3. Not all these factors carry equal weight, some being involved merely in the fine-tuning of a procedure, and the methodological details will vary for different species. Of the more

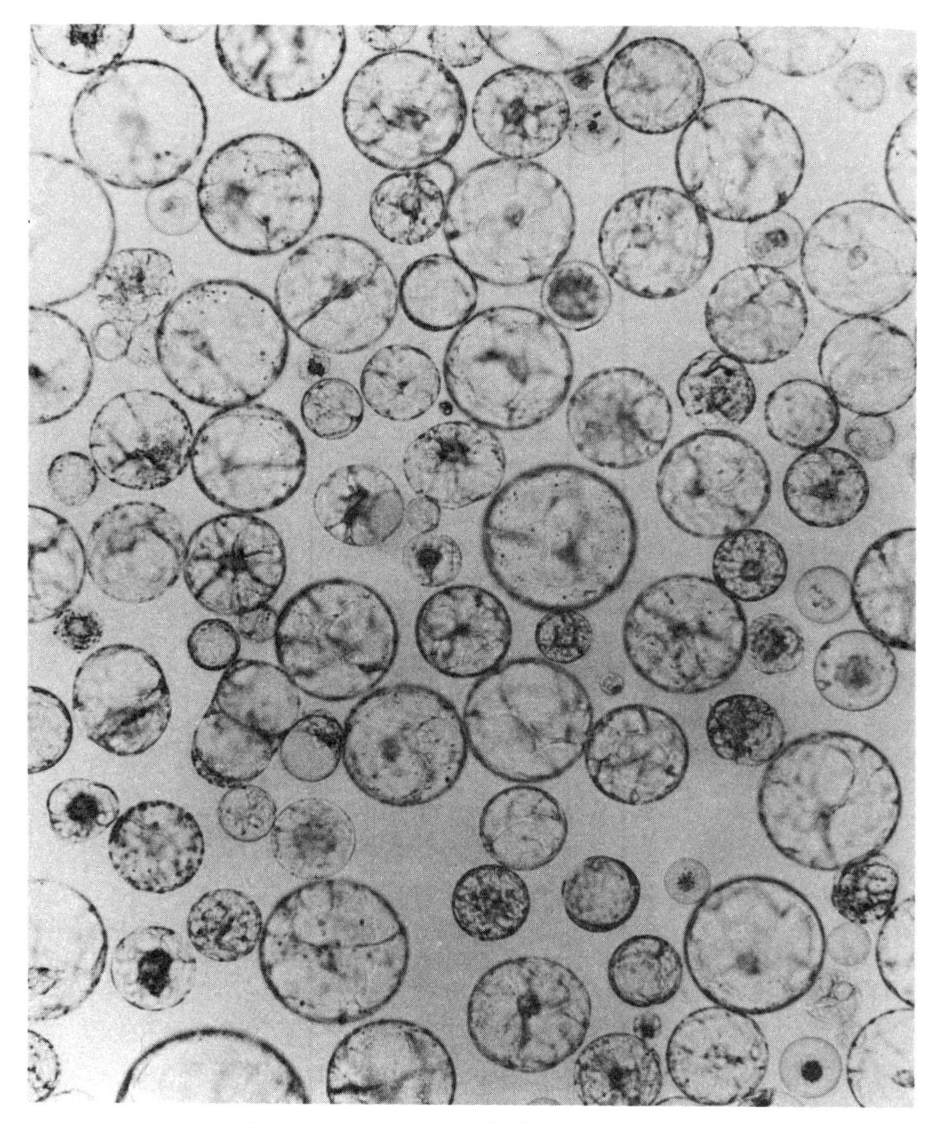

Fig. 2.10 Freshly isolated protoplasts of oilseed rape.

important factors, three in particular require serious consideration when setting up a system, being universally influential: (1) the nature of the starting material, (2) the culture medium, and (3) the plating density. Precisely what 'good' starting material is, is difficult to define. Highest yields from, for example, leaves are usually obtained from young tissues with a thin cuticle which are green rather than chlorotic, while relatively old leaves with heavily lignified tissues may be

Table 2.3 Some factors influencing protoplast yield and culture

Isolation and culture	*Digestion*	*Plant material*
• Purification procedure (e.g. sedimentation versus flotation) • Removal of residual enzyme • Composition of culture medium (e.g. osmotic pressure, growth regulators, light versus dark, temperature) • Liquid versus agarose culture	• Pre-plasmolysis • Composition of enzyme mixture • Osmoticum type, concentration • Incubation conditions in enzyme (e.g. duration, temperature, light versus dark, agitated versus stationary)	• Plant species and genotype • Source tissue: organized (e.g. leaf, embryo, hypocotyl) versus unorganized (e.g. suspension culture) • Developmental age of plant material • History of cell culture • Method used to disrupt leaf cuticle (peeling, slicing, abrasion)

difficult to digest. The precise composition of the culture medium must be defined for a particular species, and may even differ for protoplasts from different source material of the same species (e.g. leaf versus suspension culture). The concentration of the osmoticum during the digestion process should render the solution slightly hypertonic, so that the released protoplasts are slightly plasmolysed, but the osmotic potential of the culture medium is generally lowered in stages as the culture period progresses (see articles in Vasil, 1984). Finally, the plating efficiency of a culture of protoplasts, i.e. the proportion which divide, is commonly strongly influenced by the density at which they are initially cultured. Most species require a plating density within the range of 10^4 to 5×10^5 ml^{-1}, but divisions can be achieved at lower densities with the aid of nurse cultures or conditioned medium.

This chapter has served to introduce the origins and features of plant cell and tissue cultures, and in subsequent chapters it should become apparent how some of these characteristics described have been exploited for biotechnological purposes. The second group of techniques which have allowed rapid progress in plant exploitation and our understanding of differentiation and development, namely those of molecular biology, will now be discussed.

General Reading

Evans, D.A., Sharp, W.R., Ammirato, P.V. and Yamada, Y. (Eds) (1983–). *Handbook of Plant Cell Culture* (series of volumes). London & New York, Collier Macmillan.

Reinert, J. and Yeoman, M.M. (1982). *Plant Cell and Tissue Culture: a Laboratory Manual.* Berlin, Heidelberg & New York, Springer-Verlag.

Vasil, I.K. (Ed.) (1984–). *Cell Culture and Somatic Cell Genetics of Plants* (series of volumes). Orlando, Florida, Academic Press.

Yeoman, M.M. (Ed.) (1986). *Plant Cell Culture Technology.* Botanical Monographs, Vol. 23. Oxford, Blackwell Scientific Publications.

Specific Reading

Aitchison, P.A., MacLeod, A.J. and Yeoman, M.M. (1978). 'Growth patterns in tissue (callus) cultures', in *Plant Tissue and Cell Culture,* Ed. H.E. Street, pp. 267–306. Oxford, Blackwell Scientific Publications.

Fujimura, T. and Komamine, A. (1984). 'Fractionation of cultured cells', in *Cell Culture and Somatic Cell Genetics of Plants,* Vol. 1. Ed. Vasil, I.K., pp. 159–166. Orlando, Florida, Academic Press.

Hüsemann, W. (1985). 'Photoautotrophic growth of cells in culture', in*Cell Culture and Somatic Cell Genetics of Plants,* Vol. 2. Ed. Vasil, I.K., pp. 213–252. Orlando, Florida, Academic Press.

Lindsey, K. and Yeoman, M.M. (1983). 'The relationship between growth rate, differentiation and alkaloid accumulation in cell cultures', *J. Exp. Bot.* **34**, pp. 1055–1065.

Lindsey, K. and Yeoman, M.M. (1986). 'Immobilized plant cells', in *Plant Cell Culture Technology,* Botanical Monographs, Vol. 23. Ed. Yeoman, M.M., pp. 228–267. Oxford, Blackwell Scientific Publications.

Meins, F., Jr. (1986). 'Determination and morphogenetic competence in plant tissue culture', in *Plant Cell Culture Technology*, Botanical Monographs, Vol. 23. Ed. Yeoman, M.M., pp. 7–25. Oxford, Blackwell Scientific Publications.

Meins, F., Jr., Lutz, J. and Foster, R. (1980). 'Factors influencing the incidence of habituation for cytokinin of tobacco pith tissue in culture', *Planta* **150**, pp. 264–268.

Schröder, J., Betz, B. and Hahlbrock, K. (1977). 'Messenger RNA-controlled increase of phenylalanine ammonialyase activity in parsley: light-independent induction by dilution of cell suspension cultures into water', *Plant Physiol.* **60**, pp. 440–445.

Skoog, F. and Miller, C.O. (1957). 'Chemical regulation of growth and organ formation in plant tissues cultured *in vitro*', in *The Biological Action of Growth Substances*, *Symp. Soc. Exp. Biol.*, Vol. 11. Ed. Porter, H.K., pp. 118–131. Cambridge, Cambridge University Press.

Vasil, I.K. (1987). 'Developing cell and tissue culture systems for the improvement of cereal and grass crops', *J. Plant Physiol.* **128**, pp. 193–218.

Yeoman, M.M., Dyer, A.F. and Robertson, A.I. (1965). 'Growth and differentiation of plant tissues. 1. Changes accompanying the growth of explants from explants from *Helianthus tuberosus* tubers', *Ann. Bot.* **29**, pp. 265–276.

Yeoman, M.M. and Street, H.E. (1978). 'General cytology of cultured cells', in *Plant Tissue and Cell Culture*. Ed. Street, H.E., pp. 137–176. Oxford, Blackwell Scientific Publications.

Chapter 3

Molecular Biology of Plants

Plants are unique amongst all living organisms, in that the genetic material is organized in a tripartite fashion—DNA is present in the nucleus, the mitochondria and the plastids. This in itself has raised fundamental questions about the phylogeny of plants, and more directly relevant to this book, about the coordinated interactions of all three organelles in regulating gene expression during differentiation and development. As we will see later, changes in the structure and expression of specific mitochondrial and chloroplastic genes have been the subject of great scientific and commercial interest in relation to cytoplasmic male sterility and herbicide resistance, respectively, and also by virtue of the fact that within these organelles lies the machinery for energy transduction. But before we consider such applied aspects, we will provide an outline of the way in which plant genes are organized within the genome, describe strategies for the isolation and identification of a specific gene and how it may be cloned, and finally give a brief account of current knowledge of the molecular mechanisms of gene regulation.

Organization of Plant Genes

REPETITIVE SEQUENCES

The quantity of DNA in the cells of eukaryotes varies dramatically between species, and even within the plant kingdom the nuclear DNA content varies from approximately 0.5 to more than 200×10^{-12} g (Table 3.1). This does not necessarily reflect differences in the numbers of active genes in the species. Of the

Table 3.1 DNA contents of some plant species

Species	*Ploidy level*	*Unreplicated level** *of DNA (pg)*
Arabidopsis thaliana	2*x*	0.16(0.14 × 10^6kbp)†
Beta vulgaris	2*x*	2.5 (2.3 × 10^6 kbp)
Brassica napus	2*x*	3.2 (2.9 × 10^6 kbp)
Capsicum frutescens	2*x*	12.0 (11.0 × 10^6 kbp)
Hordeum vulgare cv. 'Proctor'	2*x*	10.9 (10.0 × 10^6 kbp)
Nicotiana tabacum var. *purpurea*	4*x*	15.5 (14.3 × 10^6 kbp)
Triticum aestivum cv. 'Chinese spring'	6*x*	103.8 (95.5 × 10^6 kbp)
Zea mays cv. 'Golden Bantam'	2*x*	6.5 (6.0 × 10^6 kbp)

* The unreplicated level of DNA represents the (2C) DNA content, per cell, at early interphase, i.e. after mitosis and before S phase. (Based on information in Bennett and Smith, 1976). For each species, the unreplicated level of DNA in kbp is given in brackets. 1 pg DNA ≃ 0.92 × 10^6kbp.
† More recent determination.

techniques available to study the organization of DNA, one which has provided much information on gene frequency, is that of DNA reassociation kinetics. The method exploits the fact that when double-stranded DNA is treated with alkali or high temperatures, the hydrogen bonds between pairing bases are broken and the double-helix dissociates or 'melts'; the temperature at which 50% dissociation takes place is called the melting temperature (T_m). On subsequent cooling, or on reducing the pH, complementary strands will re-anneal, and if temperature, salt concentration and DNA size (each of which affects the diffusion of DNA in solution) are controlled, the rate of reassociation of the strands is dependent on the concentration of complementary sequences and time. Theoretically, the re-association reaction should follow second-order kinetics, but in practice such a model response is not always observed, due to the probability that, with complex sequences, mismatching may occur to form a duplex which does not readily dissociate again, and correct matching is retarded. For DNA strands composed of a simple sequence of, for example, $d(A)_n{\cdot}d(T)_n$, reassociation will be relatively rapid, whereas, for plant DNA of a complex sequence, reassociation will be slower as the two complementary sequences are diluted by many other non-complementary sequences. Reassociation data are expressed in terms of a C_0t value, where C_0 represents the DNA concentration in moles of nucleotide per litre, and t represents time in seconds. Such an expression takes account of the fact that two DNA preparations of identical concentration but of different sizes will take different amounts of time to re-anneal. Based on such reassociation kinetics, measured by either UV spectroscopy or, more commonly, by hydroxyapatite chromatography (which separates single- and double-stranded DNA molecules), the plant genome can be described as being composed of three categories of sequences—highly repetitive (characterized by rapid low-concentration re-association), moderately repetitive, and low or single copy (slow, high-concentration reassociation) DNA (Fig. 3.1). However, it must be emphasized that the boundaries between these categories, and particularly between the latter

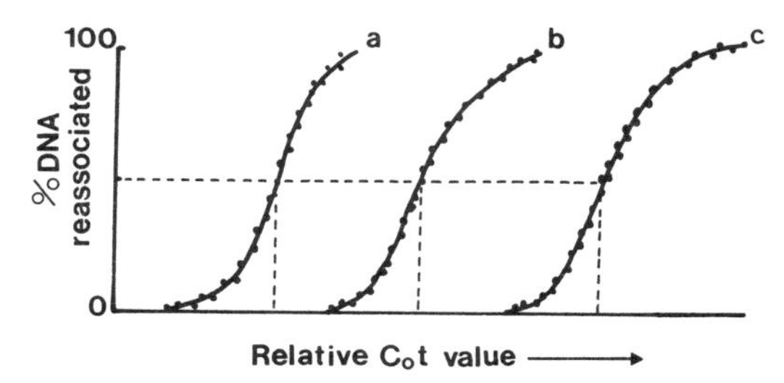

Fig. 3.1 Reassociation kinetics of DNA of differing complexity. Curves of reassociation of melted DNA of: (a) simple sequence (e.g. repetitive) DNA; (b) intermediate repetitive DNA; (c) complex sequence (e.g. unique) DNA. A 50% reassociation (-----) is observed at increasing C_0t values (i.e. increasing time for reassociation at a given DNA concentration) for DNA of increasing sequence complexity.

two, are not necessarily clearly defined. Highly repetitive DNA is also known as 'simple sequence' or 'satellite' DNA, so called because it may form a separate (satellite) band from the remainder of genomic DNA when sheared and centrifuged on a CsCl gradient; this is the result of differential buoyancy, often due to a difference in the content of adenine and thymidine residues. 'Single-copy' fractions, which are believed to contain the single or low-copy sequences representing genes encoding proteins and RNAs, comprise up to approximately 20–40% of the genome in plants, though many species may contain less than this. Species with larger genomes tend to have relatively larger proportions of highly repetitive DNA. DNA–RNA hybridization experiments indicate that not all the single-copy DNA is actively transcribed, and it is estimated that active nuclear genes represent no more than approximately 4–5% of the total genomic DNA. The function of the repetitive DNA is at present unclear, but it has been suggested that at least some sequences may have a role in the regulation of transcription. Reassociation studies have found that repetitive sequences may be located between single-copy sequences, either as short sequences of approximately the same length as single copies (approximately 1000 base pairs), as tandemly-arranged repeats of one, or of a small number of different, sequences, or as inverted or reversed repeat sequences which may form hairpin loops in the DNA, due to their palindromic structure.

The chromosomal location of repeated sequences can be determined by *in situ* hybridization, in which chromosomal preparations are hybridized with specific, radiolabelled DNA sequences and autoradiographed. Although such specific sequences may be characteristically found at certain loci, there is genetic evidence that repeated sequences may in fact be mobile and can be detected by their insertion into, and subsequent inactivation of, otherwise active genes. Such sequences have been called 'transposable elements' or 'jumping genes'. One such element, known as *Dissociator* (*Ds*), is found in maize and has been sequenced. It consists of two identical sequences of 2040 base pairs, one being inserted into the other, but in reverse orientation relative to the first; furthermore at each end of this sequence is an 11 base-pair inverted repeat. A number of other transposable elements have been identified in maize and have been sequenced, and also possess inverted repeats. Little is yet known of the role of such sequences, nor of the mechanisms by which they move around the genome; but the possibility of exploiting this phenomenon for identifying and isolating genes, and as a vector system for genetic transformation, has been recognized and will be discussed later.

Functional Genes

Functional genes can be described as being of two types—those which encode proteins, and those which encode ribosomal, transfer and smaller nuclear RNAs.

GENES ENCODING RNA MOLECULES

The RNA genes, and especially the rRNA genes (Flavell, 1986) are of particular interest, because of the huge number of ribosomes (several million) which must be synthesized per genome to maintain the protein-synthetic capability during the life cycle of the cell. The genes coding for the larger rRNA species are located in the nucleolus, and the necessarily high production of rRNA appears to be achieved by the presence of multiple gene copies which are each transcribed many times. Both the 18S and 25S rRNAs of the 80S ribosomes and the 5S rRNAs (whose genes are *not* in the nucleolus), are transcribed from tandemly-arranged repeat sequences. The precise number of the genes varies between, and even within, species, but may be up to several thousand copies per cell and representing up to approximately 1% of the nuclear DNA. The coding sequences are separated by spacer sequences of differing sizes. Transcription of the 18S and 25S sequences results in the production of polycistronic precursor molecule from which the spacer regions are excised (Fig. 3.2). Within the 5S RNA genes have been recognized two putative regulatory sequences: a consensus sequence of 5′ ATAAG 3′ is found upstream from the coding region, and is also present in *Drosophila* 5S genes, while downstream from the coding region is an AT-rich region, which may act as a signal for the termination of transcription.

GENES ENCODING PROTEINS

The protein-encoding genes are distributed in the euchromatic regions of the chromosomes, and their organization has been the subject of enormous interest for biotechnologists because of the potential for the manipulation of plant development and function through the recognition, and possible alteration, of regulatory sequences. In the simplest terms, such a gene comprises a protein coding region, upstream and downstream from which are signals for the initiation and termination, respectively, of transcription (Fig. 3.3). The coding region may possess intervening sequences (the introns) which are not represented in the final polypeptide product. On transcription of such genes, a precursor mRNA molecule is produced, from which the introns are excised ('spliced'), and the coding RNA

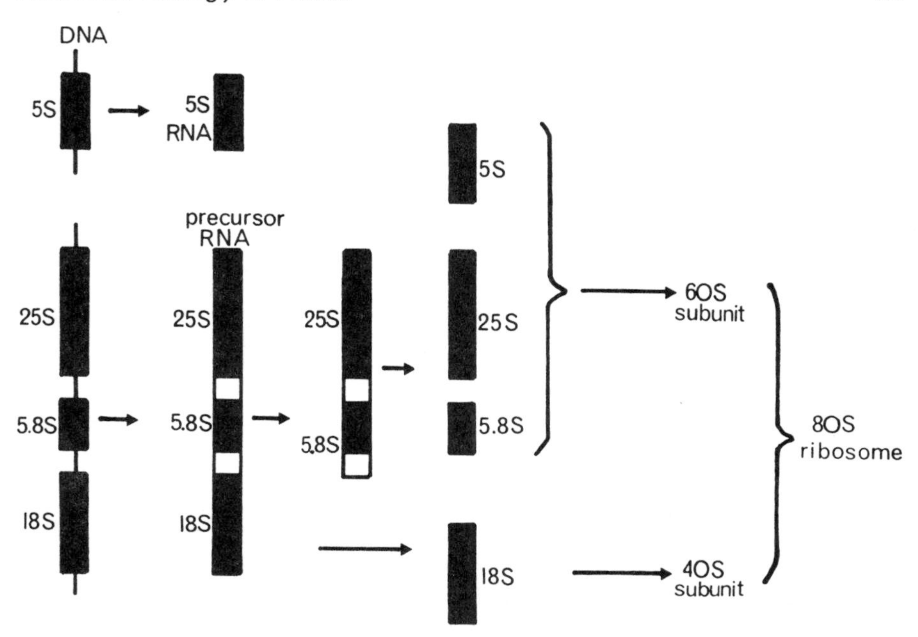

Fig. 3.2 Organization and splicing of rRNA genes and transcripts. The 18S, 5.8S and 25S rRNA genes are transcribed to produce a polycistronic precursor molecule which is methylated and spliced. The 5.8S rRNA is derived from part of the 25S rRNA, and the 5S is encoded at a separate locus.

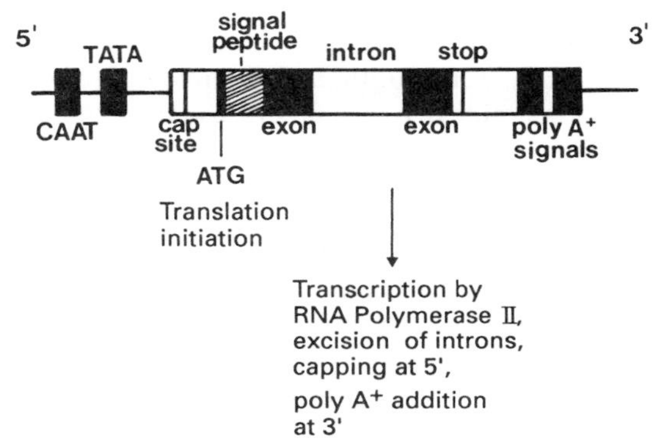

Fig. 3.3 Organization of a hypothetical gene encoding a plant protein.

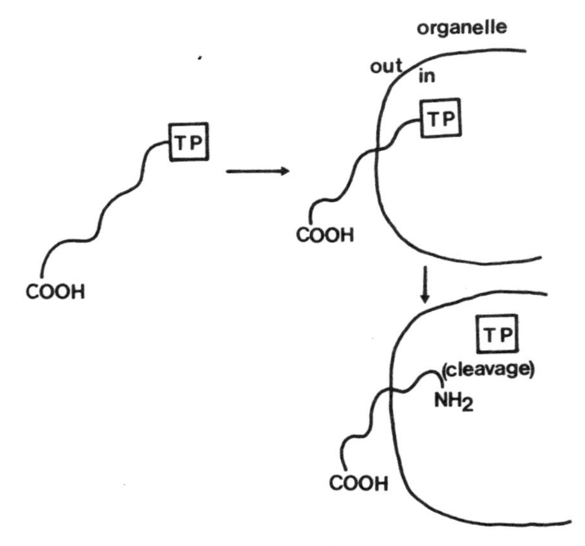

Fig. 3.4 Diagrammatic representation of the role of a transit peptide (TP). The TP directs the protein to a specific organelle, and is cleaved during membrane transport to leave the mature protein *in situ*.

fragments (the exons) are ligated together prior to translation. The function of introns is unclear, but in yeast, one intron in the mitochondrial gene for cytochrome oxidase subunit-I has been shown to encode a 'maturase' which itself acts to splice the transcripts (Carignani *et al.*, 1983). The presence of introns has also been shown to influence strongly the expression of genes (e.g. alcohol dehydrogenase) in transiently and stably transformed protoplasts (Callis *et al.*, 1987). Intron sequences are therefore not merely a relict of evolution, but may be of functional significance.

In some genes there have been found sequences which encode a peptide which is found at the N-terminal region of the translation product, but which is excised before the mature protein is assembled. Such peptides, usually less than approximately 60 amino acids in length, have been found to be active in directing the transport of the polypeptide to a specific site, usually a membranous organelle, in the cell. These so-called 'transit peptides' are excised from the polypeptide during or after membrane transport at the target organelle (Fig. 3.4), and it has been shown to be possible to target marker genes linked to chloroplast- and mitochondrion-specific peptides (e.g. Boutry *et al.*, 1987). The possible biotechnological value of the phenomenon will be discussed later.

Both 5′ and 3′ to the coding region of the protein genes are sequences which are highly conserved (consensus sequences) which appear to be involved in the initiation and termination of transcription. Upstream to the ATG codon (representing the start of translation), the so-called 'CAAT' and 'TATA' boxes are found in many, but not all, plant and animal genes, although the precise

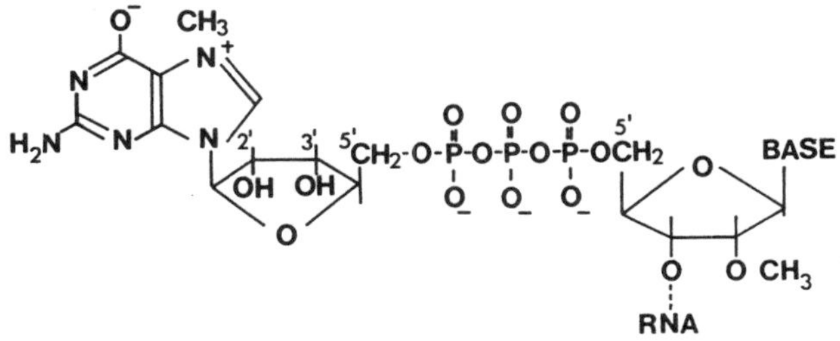

Fig. 3.5 The structure of 7-methyl guanosine.

nucleotide sequence is variable. The 'TATA' box may determine the site of transcription initiation by acting as a binding site for RNA polymerase-II, while the 'CAAT' box may act as an enhancer element in modifying transcription in a quantitative way. The functional significance of such sequences has been confirmed by both deletion mutation experiments and by linking them to marker genes and following their expression after introduction into protoplasts, or in transgenic plants (see later).

The 5′ ends of the transcripts (i.e. pre-mRNAs) of at least some plant genes are characterized by the presence of methylated nucleosides, notably 7-methyl-guanosine, attached by a triphosphate linkage after transcription (Fig. 3.5). There is evidence that the spatial relationship between this so-called CAP site and the ATG is important in determining the level of expression of the gene, and the CAP itself may act as a binding site for a regulator of translation initiation or in stabilizing the message. Interestingly, chloroplast and mitochondrial mRNAs appear to have no CAP structure.

The 3′ ends of many (probably all) plant mRNAs are also post-transcriptionally modified by the addition of polyadenylic acid or poly(A) sequences; such transcripts are designated as poly(A^+) RNA, and are isolated by binding to oligo(dT) cellulose or poly(U) sepharose. Although the poly(A) is added after transcription and the sequence is not itself encoded, a signal for the attachment does reside in the downstream region of the gene. Usually, poly(A) addition (up to 150–200 nucleotides), mediated by poly(A) polymerase, occurs on transcripts possessing a 3′ GC followed, approximately 20 bases later, by an AAUAA sequence. The function of the poly(A) sequence is related to the stability of the transcript, and much evidence for this view is derived from animal work. In general, the size of the poly(A) sequence is reduced during the lifetime of a transcript; and poly(A^-) transcripts may have shorter half-lives than poly(A^+) RNAs. The presence of a poly(A) sequence appears not, in itself, to be essential for the successful translation of the transcript.

MULTIGENE FAMILIES

There is evidence that there exist in the genomes of plants and animals families of structurally-related genes, perhaps numbering up to 100 individuals, which encode similar proteins. The products of the so-called 'multigene families' may have similar physical and biochemical properties, and include storage proteins of cereals (e.g. the hordeins of barley, the gliadins of wheat) and potato (patatin), proteinase inhibitors (e.g. the chymotrypsin inhibitors of barley) and isoenzymes. However, not all the members of a multigene family are necessarily transcribed, and some of these may be described as 'pseudogenes'. Some pseudogenes may lack the introns found in transcribed members of the same family, or may possess modified intervening sequences (perhaps as the result of transposition by mobile elements).

Gene Isolation and Identification

Plant genetic engineering has become a rapidly expanding field since the development of methods for isolating and identifying specific genes. The available strategies can be described as falling into one of four categories:

(1) the production and screening of cDNA and genomic libraries;
(2) the study of restriction fragment length polymorphisms (RFLPs);
(3) transposon and insertional mutagenesis; and
(4) the transfer of DNA fragments of unknown genetic constitution to *E. coli*, yeast or plant cells, to study possible phenotypic or biochemical changes ('shotgun cloning').

cDNA AND GENOMIC LIBRARIES

The screening of genomic libraries (also called 'gene banks') has proved to be a very powerful technique for the isolation of genes encoding tissue-specific, developmentally-specific, metabolic and storage proteins in a range of plant species. The essential features of the construction of genomic libraries are summarized in Fig. 3.6, and are: (1) the generation of fragments of DNA, which are (2) introduced into a cloning vector such as a virus, a plasmid or a cosmid (discussed below), which is (3) itself introduced and allowed to replicate in a bacterial cell such as *E. coli*. The culture of the bacteria results in the production of a large number of clones of the DNA fragments, which can be screened for a particular gene of interest (see below). By cutting extracted genomic DNA with two restriction enzymes, a library of random fragments can be produced, and Clarke and Carbon (1976) have shown that it is possible to calculate the probability of including any particular DNA sequence in such a library. For example, in a library made from the human genome, which is 2.8×10^6 kilobases in size and comprising fragments of 20 kilobase pairs, it is necessary to screen 4.2×10^5 recombinants to detect a specific sequence with a probability of success of 95%. It is possible to increase the chances of successful screening by enriching fractions of DNA fragments using chromatographic techniques, but if a gene product is believed to be abundant at a particular stage of development, or in a particular

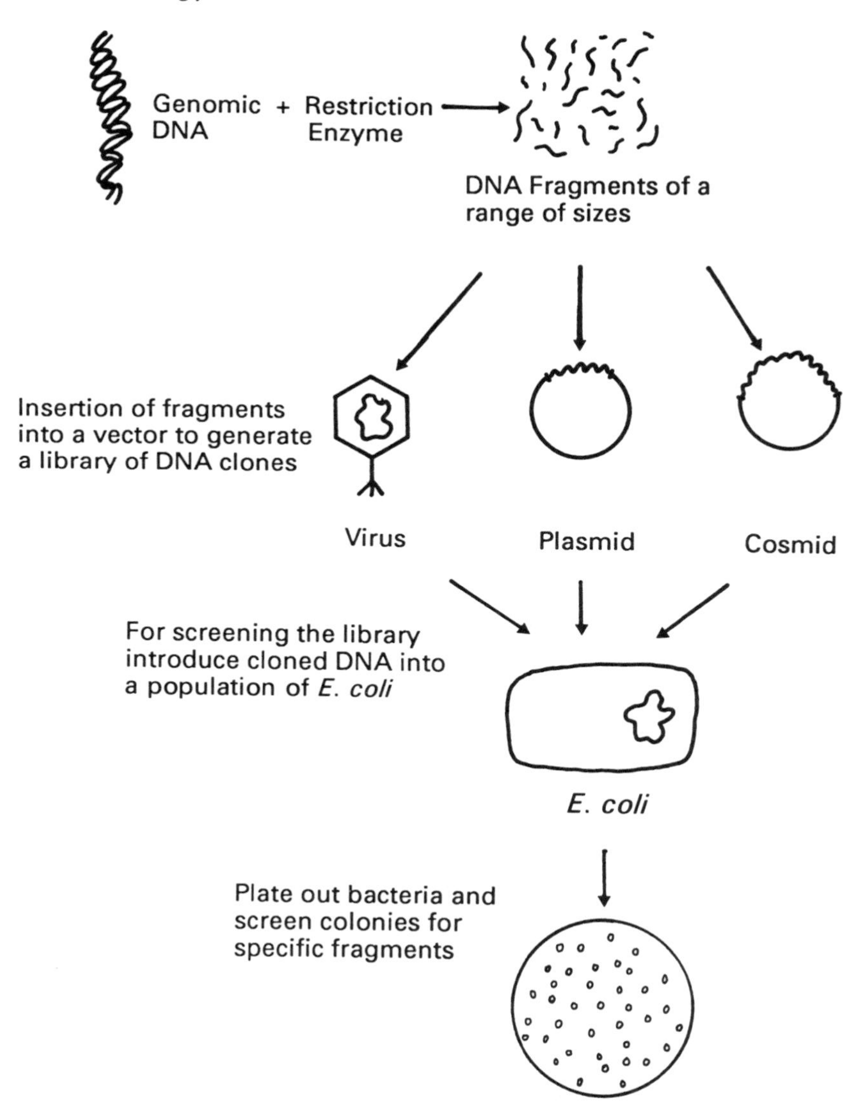

Fig. 3.6 Construction of a genomic library. DNA fragments are cloned in either virus particles, plasmids or cosmids according to their size. Plasmids are used for relatively small fragments only.

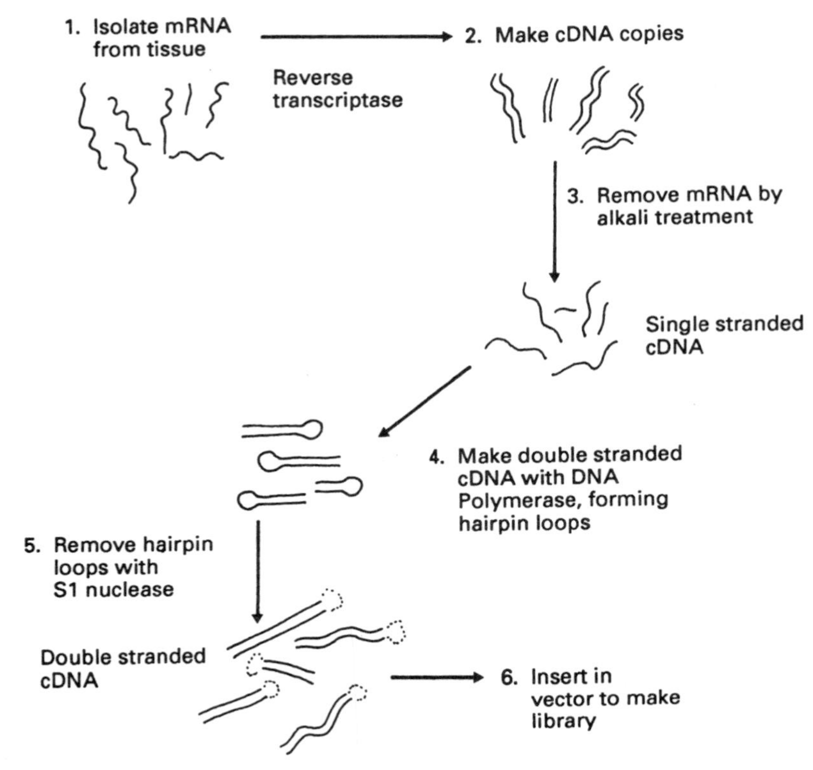

Fig. 3.7 Construction of a cDNA library. Poly(A^+) RNA is purified on oligo(dT) cellulose or poly(U) sepharose, and synthetic oligo(dT) is added to prime the cDNA synthesis reaction.

tissue, then an alternative strategy is to produce a cDNA library (Fig. 3.7). In this technique, mRNAs are isolated and the enzyme reverse transcriptase, which naturally occurs in retroviruses and catalyses the production of a DNA strand from an RNA template, is used to generate complementary DNA (cDNA) copies of the transcripts. Cloning of the cDNAs therefore produces a library of those genes which are actively transcribed, and highly or moderately repetitive sequences, which are transcriptionally inactive, are not represented. As a result, the probability of successfully recovering a gene of interest may be greatly increased.

Detection of a Gene within a Library

A variety of methods is currently available to recognize specific genes within a library. If a purified protein is available, the amino acid sequence can be determined and it may be possible to chemically synthesize an oligonucleotide (usually at least 15–20 nucleotides long) which is colinear to a unique region of the

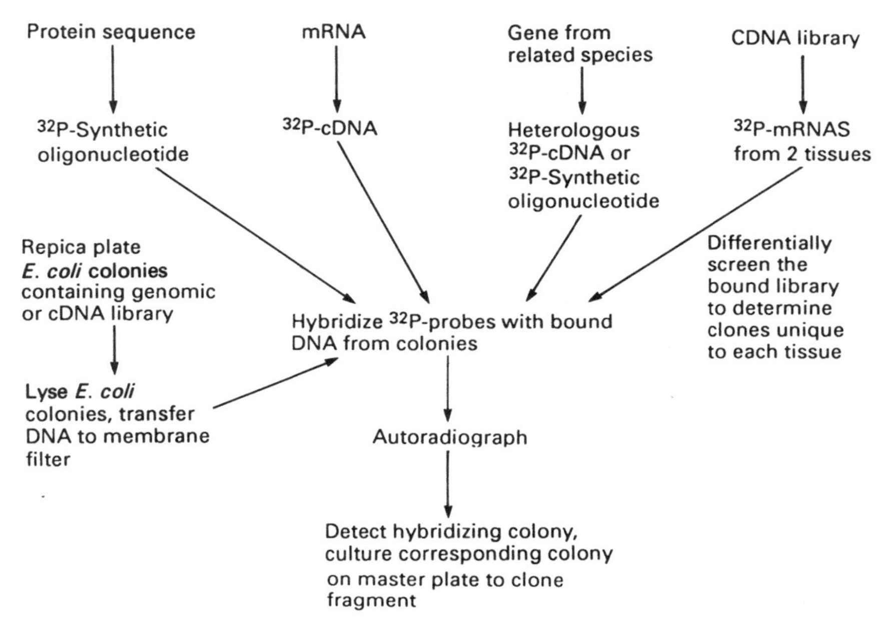

Fig. 3.8 Use of radioactive probes to identify genes within a library.

polypeptide. Such a sequence will hybridize to single-stranded DNA sequences in the library (after denaturation of the cloned fragments) if the gene (or part of the gene) encoding that particular amino acid sequence is represented (Fig. 3.8). A potential problem is the degeneracy of the genetic code—most amino acids are encoded by more than one nucleotide triplet, which could lead to mismatching of the synthetic probe. This problem can be solved in part by synthesizing oligonucleotides complementary to regions of the amino acid sequences which contain methionine and tryptophan residues, which are encoded by single codons, and it is also possible to use mixed populations of oligonucleotides to cover the various permutations.

Analysis of the sequences of a number of individual genes isolated from a range of species has shown the presence of consensus sequences—i.e. a region of common primary structure which has been conserved, despite the evolutionary divergence of the species. The sequence similarity may be such that it may be possible to use a radioactive nucleotide sequence complementary to a gene isolated from one species to screen the library of a second species, and so isolate the gene for a related (but not necessarily identical) protein. This technique is particularly useful for isolating genes for metabolically essential proteins, the structures of which are likely to be conserved by strong selection pressure. For example, the gene for the mitochondrion-encoded subunit-II of cytochrome oxidase in maize has been isolated using a 'heterologous' probe from yeast.

It is also possible to isolate a specific transcript from extracted total or poly (A^+) RNA by virtue of its ability to hybridize to a cDNA clone. After hybridization to an immobilized cDNA, unbound RNA can be removed by washing, the remaining RNA strand can be recovered and translated in a cell-free system ('hybrid-select' or 'hybrid-released translation'), and the polypeptide can be subsequently analysed. This technique was, for example, used to identify B hordein genes in cDNA libraries of barley. Alternatively, hybrid *arrested* translation can similarly be used to identify specific transcripts, but relies on the fact that mRNA is not capable of translation when bound to a complementary DNA sequence. Thus, when translation products, separated by electrophoresis, are compared from samples prepared before and after hybridization, it is possible to detect that polypeptide which is encoded by the bound mRNA (Fig. 3.9).

cDNAs are synthesized from mature mRNA molecules, in which introns and 5′ and 3′ regulatory sequences are absent. For the genetic manipulation of a specific agronomic trait it may be required to transfer a full-length gene into a host plant, and cDNA clones, while themselves incomplete sequences, can nevertheless be used to probe genomic libraries to retrieve unspliced sequences. Such a method can be used to isolate genomic clones of genes which are expressed in a developmentally-regulated or tissue-specific fashion, such as those for cereal storage proteins or fruit-ripening-specific genes (see Chapter 8).

One other method for isolating, for example, development- or tissue-specific genes, is that of differential hybridization or 'plus-and-minus' screening. Here, cDNA libraries are made from two tissues. One library is probed with both the radioactively-labelled libraries (or even the mRNA) independently, and clones which are common to both tissues will hybridize to both. The interesting clones, specific to one of the tissues, will not, however, hybridize to both libraries, and can be cloned for further investigation and used to probe a genomic library.

RESTRICTION FRAGMENT LENGTH POLYMORPHISMS (RFLPS)

Many agronomic traits, such as resistance to fungal attack, or qualitative features such as fruit texture or taste, are not encoded by single genes but are 'polygenic'—determined by the interaction of two or more gene products. The genes involved may be localized far apart on a single chromosome or on separate chromosomes, and because of the difficulties in recognizing the specific proteins involved in complex metabolic processes it is not possible to obtain gene probes using the methods so far described. An alternative approach is to construct linkage maps of the genome using, as genetic markers, restriction fragment length polymorphisms (RFLPs). If genomic DNA from an individual plant is digested with a single restriction endonuclease, and the fragments are separated by gel electrophoresis, a characteristic banding pattern will be produced on probing with a radioactive DNA clone; the precise genetic nature of the clone does not necessarily need to be known. If mutations are present in the same restriction site in DNA from a second individual, then a different restriction pattern will be produced by virtue of the differences in restriction fragment lengths (see Chapter 5, Figs. 5.5 and 5.6). It is

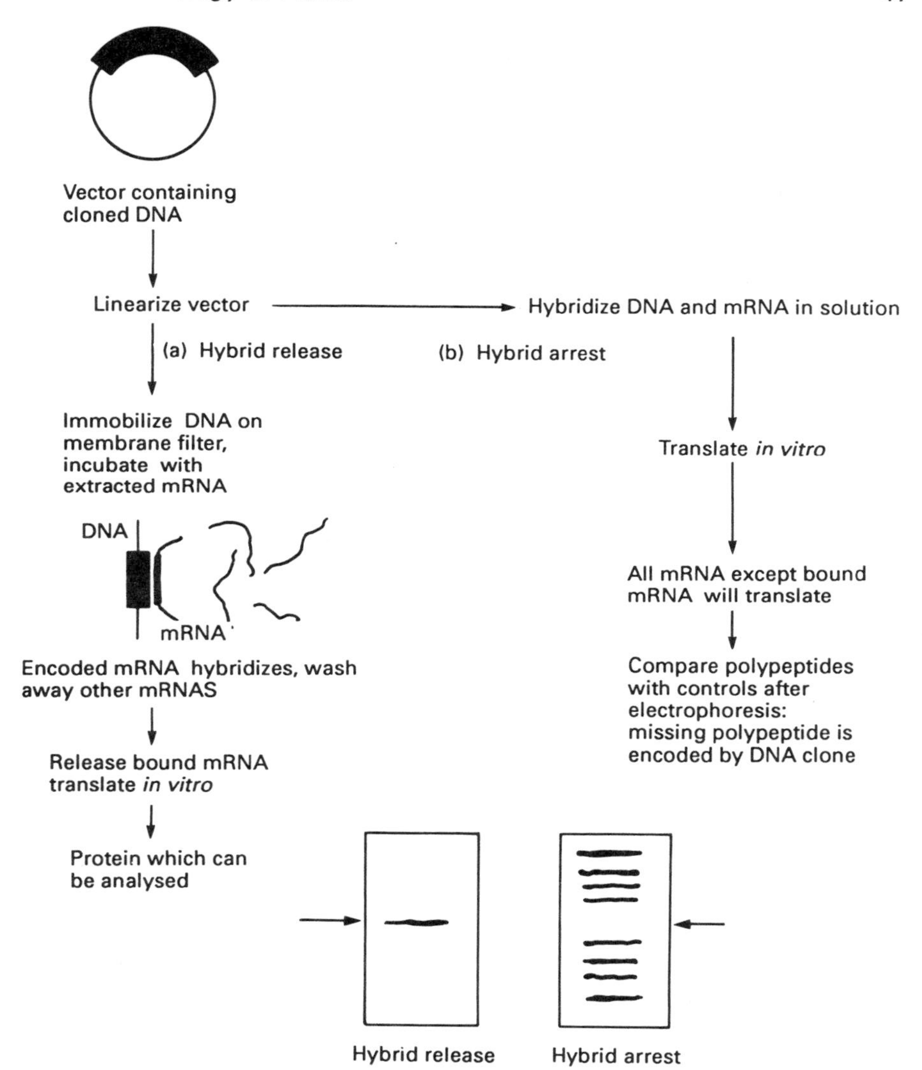

Fig. 3.9 Hybrid release and hybrid arrest translation to identify specific mRNAs.

possible to create linkage maps of RFLP loci by analysing the changes in restriction patterns of DNA extracted and analysed after a series of reproductive crosses. It is also possible to correlate RFLP loci with the presence of a particular polygenic trait, and RFLP linkage has been observed with, for example, the genes controlling the soluble solids content of tomato fruit (Osborn *et al.*, 1987). It is hoped that the recognition of such correlated RFLPs will lead to the identification

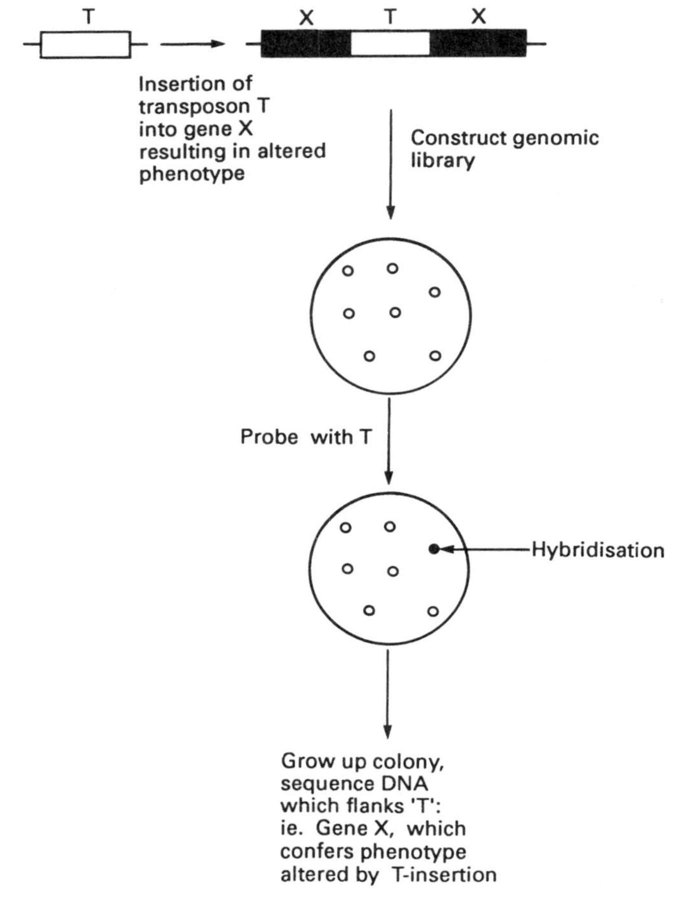

Fig. 3.10 Transposon tagging of genes.

of fragments encoding groups of genes involved in a particular trait. It has to be cautioned, however, that the technique may only be of value to follow the segregation of relatively close-linked genes.

TRANSPOSON AND INSERTIONAL MUTAGENESIS

A third strategy which has been used with success, and can be expected to become more widely adopted as transformation techniques for plants are improved, is transposon and insertional mutagenesis. The rationale of the technique is the structural and functional alteration or inactivation of a gene which occurs when a foreign DNA sequence introduced into the plant is inserted into it. Such a situation is analogous to the action of naturally-occurring transposable elements

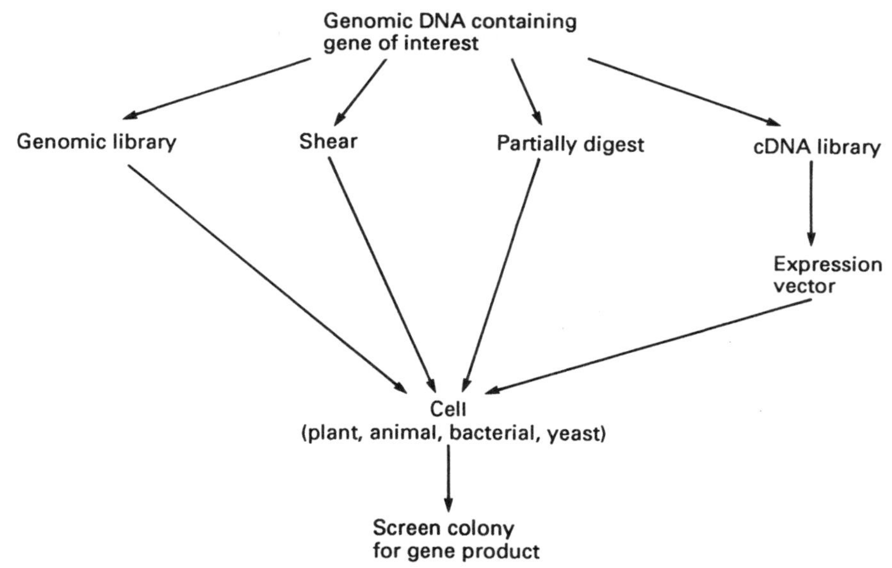

Fig. 3.11 General strategy of 'shotgun cloning'.

(see above). If the introduction of such a transposon results in an altered phenotype, then a radioactively-labelled DNA sequence complementary to the transposon can be used to probe DNA fragments in a genomic library prepared from transformed plants, so that gene sequences flanking the transposon can be isolated (Fig. 3.10). This method is likely to be of little value for the tagging of genes encoding polygenic traits, due to the low probability of the simultaneous insertional mutagenesis of two or more separate loci. Examples of genes which have been identified by this technique include those on the Ti plasmid of *Agrobacterium* (see below), and on *Rhizobium* plasmids.

Chapter 10 discusses the prospects of using transposon mutagenesis to study the regulation of genes and of generating novel regulatory sequences.

THE 'SHOTGUN' APPROACH

The fourth method we will consider for isolating genes is described as a 'shotgun' technique, because it involves the introduction of random DNA fragments into cells of *E. coli*, yeast, plants or animals, followed by analysis of the gene products or of the effect on the phenotype of the cell. The general strategy can also be used to study the function of specific genes, the translation products of which are unknown, or of engineered genes. The introduced DNA may be 'naked' (i.e. simply sheared or partially digested by one or two restriction enzymes) with expression under the control of its own regulatory sequences, or cDNAs may be incorporated into an expression vector cassette, possessing alternative (e.g. constitutive) promoter and termination/polyadenylation sequences (Fig. 3.11).

This general method has been used to study, for example, proteins encoded by the Ti plasmid, and expressed in *E. coli*, and can be used for the isolation of selectable genes. For example, Klee *et al.* (1987) of the Monsanto company have developed a shotgun cloning technique to detect, and subsequently isolate, a mutant gene from *Arabidopsis thaliana*, encoding an acetolactate synthase (ALS) enzyme insensitive to sulphonylurea herbicides (see Chapter 9). A cosmid library of the *Arabidopsis* genome was introduced into *Agrobacterium tumefaciens*, which in turn was used to inoculate *Petunia hybrida* leaf discs. Plants regenerated in the presence of chlorsulphuron were transformed with the mutant ALS gene, demonstrating the successful transfer of a valuable trait without the need to isolate and clone a specific gene. Potentially the technique could be exploited to look for genes regulating disease resistance or developmental processes.

Cloning Strategies

Once a specific DNA fragment has been detected, further analysis requires it to be produced, *in vitro*, in relatively large quantities. We will now consider briefly how this may be achieved.

Methods for the cloning of specific DNA sequences all involve inserting those sequences into 'vector' DNA which is capable of replicating within a host cell (e.g. *E. coli*) and which carries a selectable gene, usually encoding resistance to one or more antibiotics, so ensuring the preferential propagation of the desired gene construct. In *E. coli*, three types of cloning vectors have been developed: plasmids, bacteriophages and cosmids.

PLASMIDS

Plasmids are double-stranded circles of DNA, found naturally in a range of prokaryotes and in some eukaryotic cells (e.g. the S plasmids of cytoplasmically male-sterile maize; see Chapter 8), and are capable of replicating independently of the chromosomal DNA. Plasmid genes may encode a variety of traits, including resistance to antibiotics and heavy metals, the ability to ferment sugars, and, in relation to interactions with plants, genes for tumour or root induction (*Agrobacterium tumefaciens*, *A. rhizogenes*) and for nitrogen fixation (*Rhizobium*). Amongst the most commonly-used plasmids for use in *E. coli* are those based on pBR322, notably the pUC series, which have been synthesized specifically for cloning purposes and which have a number of unique restriction sites which allow the ready insertion of foreign sequences. Cloning in plasmids is most efficient when the size of the insert is relatively small, usually no more than 5 or 6 kilobase pairs.

BACTERIOPHAGES

For the cloning of larger fragments, bacteriophage vectors, notably λ, have been developed. Bacteriophage λ is a virus which attacks *E. coli*, and comprises a genome of 50 kilobase pairs of linear double-stranded DNA which is packaged into a protein coat. At the termini of the DNA are complementary single-stranded

5′ sequences which join together in the host cell to form the *cos* (cohesive) site. Although the wild-type λ phage is unsuitable as a cloning vector, because of the presence of multiple sites for each of a number of commonly-used restriction enzymes, deletion modifications have been made which allow the packaging, at unique cloning sites, of up to 20 kilobase pairs of foreign DNA.

COSMIDS

Further removal of λDNA has revealed that only a relatively small sequence, close to the *cos* site, is required for recognition and successful packaging of DNA into the phage particles. By inserting this *cos* region into a plasmid, e.g. of size approximately 5 kilobase pairs, it has been possible to package up to about 45–50 kilobase pairs of foreign DNA into the virus coat, and subsequently become propagated, in a circular form, after transduction of *E. coli*. Such hybrid vectors are termed 'cosmids'.

Finally, it is worthy of mention that cloning vectors such as phage M13 have been developed which have the advantage that single-stranded clones can be produced which can be readily sequenced or used to produce cDNA strands. For a more detailed treatise of this and the other cloning vectors briefly described, and also of sequencing techniques, the reader is referred to Maniatis *et al.* (1982) and Old and Primrose (1985).

Regulation of Gene Expression

To summarize the principal features of current knowledge which relates plant gene structure to function, we have already seen how specific sequences outside of the protein coding region are conserved between genes. Recent developments in plant transformation techniques have now been used to demonstrate unequivocally the roles of 5′ and 3′ sequences in regulating transcription in response to tissue-specific, development-specific and environmentally-induced factors. For example, cereal seed storage proteins such as the barley hordeins and wheat glutenins are synthesized and accumulated solely in the starchy endosperm, and by linking a 500 base pairs upstream regions of the encoding genes to bacterial marker genes such as chloramphenicol acetyltransferase (CAT), it has been demonstrated that 5′ sequences direct endosperm-specific expression in transgenic tobacco plants (see Chapter 8). Similar techniques have been used to study and demonstrate root- and tuber-specific synthesis of the protein patatin in transgenic potato plants, and the inducibility of stress-related genes, such as proteinase inhibitors (on tissue wounding) and alcohol dehydrogenase (under conditions of anaerobiosis). In such cases, such specificity can be overcome by replacing these so-called *cis*-acting upstream sequences with a constitutively active promoter, such as the cauliflower mosaic virus 35S RNA gene promoter (see review by Kuhlemeier *et al.*, 1987). Furthermore, a combination of tissue-specific and environmentally-inducible promoters, artificially fused together, have been shown to generate a unique pattern of gene expression in transgenic plants.

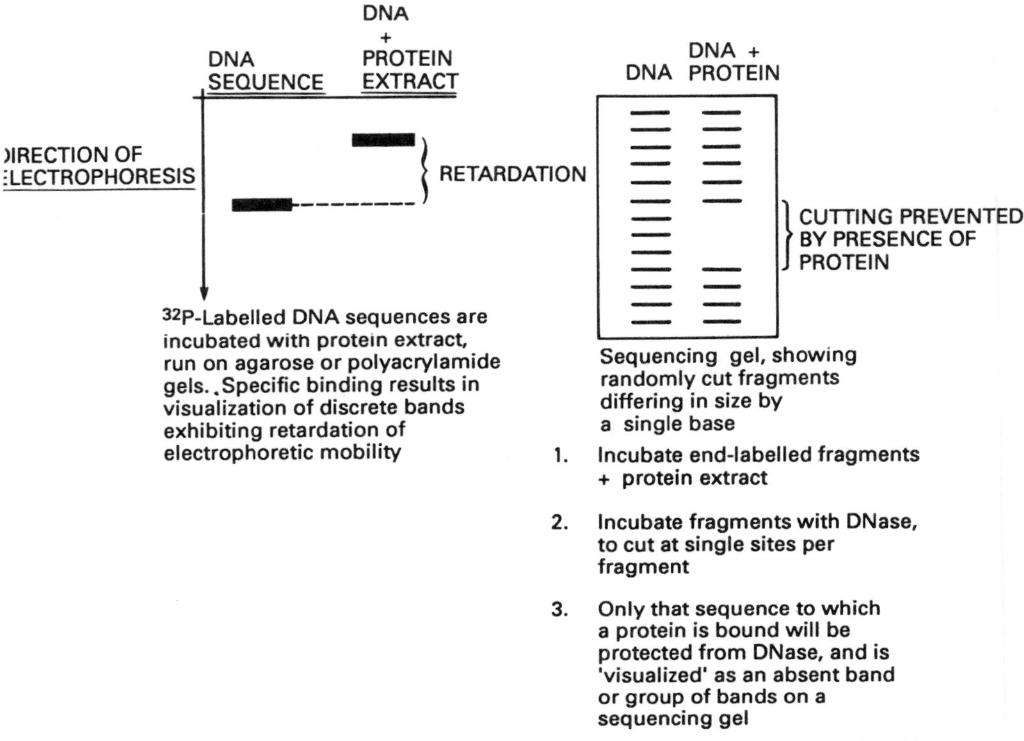

Fig. 3.12 General methods to detect DNA sequences to which proteins bind. Left, retardation gels; right, DNAse protection.

Since it is believed that all cells of an organism contain identical copies of all genes, the specificity of transcription which has been observed must be orchestrated by soluble, *trans*-acting factors, and much interest is currently directed to studies on DNA-binding proteins, the products of presumed 'regulatory genes'. It can be hypothesized that such factors may act to increase (positive control) or decrease (negative control) the rate of transcription of specific genes, and a summary of two approaches to examine such interactions is illustrated in Fig. 3.12. The first method relies on the fact that the electrophoretic movement of a specific DNA fragment will be retarded if it is bound to a protein, and the second exploits the protection of a DNA fragment by a protein to the enzyme DNase. Using such techniques, it has been possible to detect proteins which bind to regulatory DNA sequences (e.g. Jofuku *et al.*, 1987), but such work is at an early stage. For excellent reviews of *cis*- and *trans*-regulatory factors, the reader is referred to Leaver *et al.* (1986) and Kuhlemeier *et al.* (1987). Specific regulatory mechanisms for individual genes are discussed in subsequent chapters, where appropriate.

DNA Methylation

There is increasing evidence that the post-replicative covalent modification of DNA, and specifically the methylation of bases, may play a role in the regulation of gene expression in both prokaryotes and eukaryotes (Doerfler, 1983). It has been known for many years that most restriction enzymes will not cleave at their normal sites if one or more bases at that site is methylated, although a small number of enzymes, such as *Dpn*I (from *Diplococcus pneumoniae*), will only cut if methylation has occurred. Roles for methylation in protection against restriction enzymes (e.g. of viruses) and in DNA replication and repair have been established in prokaryotes, and in addition, studies with inhibitors of methylation, such as 5-azacytidine, and with the introduction of covalently-modified genes into mammalian cells by transfection, have provided data to implicate methylation in modifying the transcriptional activity of specific eukaryotic genes. Prokaryotic DNA contains the modified bases N^6-methyladenine (6-mA) and 5-methylcytosine (5-mC), and while some green algae, protozoa and insects may possess very low amounts of 6-mA, 5-mC is much more common, and is the only modified base found in significant levels in the higher eukaryotes. 5-Azacytidine, an analogue of 5-mC which can become integrated in DNA, prevents methylation of DNA predominately by its ability to trap methyltransferases and thereby inhibit their activity. Treatment of plant and animal cells with 5-azacytidine has resulted in the activation of previously silent genes, and recently a strong inverse correlation has been established between the degree of methylation and the tissue-specific expression of seed storage proteins in maize (Bianchi and Viotti, 1988). Here, zein and glutelin genes were found to be heavily methylated in DNA extracted from pollen and somatic tissues, in which there is no expression, whereas in the DNA of endosperm analyzed 22 days after pollination there was significant under-methylation, as determined by changes in restriction patterns after digestion of the DNA with methylation-sensitive restriction enzymes. These are the first data to link methylation with tissue-specific gene expression in plants.

The mechanisms of the action of methylation in modifying gene expression are not yet clear, but probably the most significant effect in relation to this is the corresponding alteration in the organization of chromatin. It is known that DNA can assume a number of configurations other than the classical right-handed 'B-form', and it has been shown that the transition to the left-handed 'Z-form' can be facilitated and stabilized by 5-mC or 7-methylguanine. The significance of such changes in DNA structure in relation to transcription has not been defined. Furthermore, with increasing evidence for the importance of DNA–protein interactions in modulating transcription, it might be expected that methylation could exert its effect at this level of regulation by modifying those interactions. Future studies in which specific promoter or termination sequences are modified by methylation should produce further relevant information.

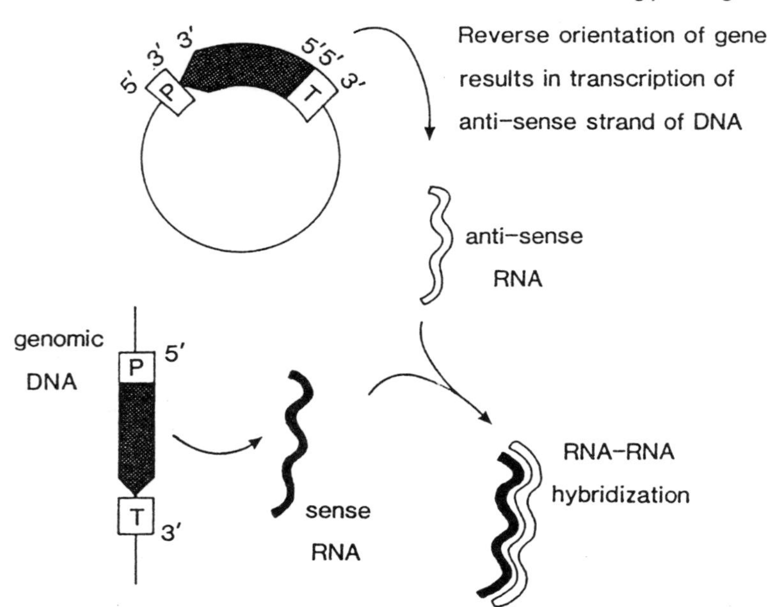

Fig. 3.13 Summary of the synthesis and action of anti-sense RNA.

Anti-sense RNA

Anti-sense RNA can be defined as RNA which binds to messenger (or sense) RNA to form a double-stranded structure incapable of being translated on ribosomes. There is now evidence that the expressions of some prokaryotic genes, such as the thymidine kinase gene of *E. coli*, are regulated by an anti-sense mechanism, but as yet the natural occurrence of such a system in eukaryotic cells has not been demonstrated. Nevertheless, the use of artificially synthesized anti-sense RNA has been proposed as a means by which the experimental regulation of plant and animal genes can be manipulated, and some success has been achieved.

Anti-sense RNA can be synthesized by the general method illustrated in Fig. 3.13. If the coding region of a gene is fused to its promoter in reverse orientation, the anti-sense strand of the DNA will be transcribed to produce an anti-sense RNA. Transformation of plant and animal cells with anti-sense genes, or the direct introduction of anti-sense RNA by injection, has been used specifically to block the translation of the mRNA corresponding to the genes, such as those encoding heat shock proteins, enzymes involved in anthocyanin biosynthesis in flower pigmentation or polygalacturonase, an enzyme involved in tomato fruit wall softening. Some prospects for using this strategy to study plant gene expression are discussed by Verma *et al.* (1987).

Radioactively-labelled anti-sense RNA has been used to localize mRNAs *in situ*, by hybridization with tissue sections (e.g. Martineau and Taylor, 1986). This technique has much potential for elucidating tissue-specific and development-specific gene expression in plants.

General Reading

Grierson, D. and Covey, S.N. (1984). *Plant Molecular Biology*. Glasgow, London, Blackie.

Leaver, C.J., Boulter, D. and Flavell, R.B. (Eds) (1986). 'Differential gene expression and plant development', *Phil. Trans. R. Soc. Lond.* **B314**. London, The Royal Society.

Maniatis, T., Fritsch, E. and Sambrook, J. (1982). *Molecular Cloning: a Laboratory Manual*. New York, Cold Spring Harbor Laboratory.

Old, R.W. and Primrose, S.B. (1985). *Principles of Gene Manipulation: an Introduction to Genetic Engineering*, 3rd ed. Oxford, Blackwell Scientific Publications.

Specific Reading

Bennett, M.D. and Smith, J.B. (1976). 'Nuclear DNA amounts in angiosperms', *Phil. Trans. R. Soc. Lond.* **B274**, pp. 227–274.

Bianchi, M.W. and Viotti, A. (1988). 'DNA methylation and tissue-specific transcription of the storage protein genes of maize', *Plant Mol. Biol.* **11**, pp. 203–214.

Boutry, M., Nagy, F., Poulsen, C., Aoyagi, K. and Chua, N.-H. (1987). 'Targeting of bacterial chloramphenicol acetyltransferase to mitochondria in transgenic plants', *Nature* **328**, pp. 340–342.

Callis, J., Fromm, M. and Walbot, V. (1987). 'Introns increase gene expression in cultured maize cells', *Genes Devel.* **1**, pp. 1183–1200.

Carignani, G., Groudinsky, O., Frezza, D., Schiavon, E., Bergatino, E. and Slonimski, P.P. (1983). 'An mRNA maturase is encoded by the first intron of the mitochondrial gene for the subunit 1 of cytochrome oxidase in *S. cerevisiae*', *Cell* **35**, pp. 733–742.

Clarke, L. and Carbon, J. (1976). 'A colony bank containing Col E1 hybrid plasmids representative of the entire *E. coli* genome', *Cell* **9**, pp. 91–99.

Doerfler, W. (1983). 'DNA methylation and gene activity', *Ann. Rev. Biochem.* **52**, pp. 93–124.

Flavell, R.B. (1986). 'The structure and control of expression of ribosomal RNA genes', *Oxf. Surv. Plant Mol. Cell Biol.* **3**, pp. 251–274.

Jofuku, K.D., Okamuro, J.K. and Goldberg, R.B. (1987). 'Interaction of an embryo DNA binding protein with a soybean lectin gene upstream region', *Nature* **328**, pp. 734–737.

Klee, H.J., Hayford, M.B. and Rogers, S.G. (1987). 'Gene rescue in plants: a model system for "shotgun" cloning by retransformation', *Mol. Gen. Genet.* **210**, pp. 282–287.

Kuhlemeier, C., Green, P.J. and Chua, N.-H. (1987). 'Regulation of gene expression in higher plants', *Ann. Rev. Plant Physiol.* **38**, pp. 221–257.

Martineau, B. and Taylor, W.C. (1986). 'Cell-specific photosynthetic gene expression in maize determined using cell separation techniques and hydribization *in situ*', *Plant Physiol.* **82**, pp. 613–618.

Osborn, T.C., Alexander, D.C. and Fobes, J.F. (1987). 'Identification of restriction fragment length polymorphisms linked to genes controlling soluble solids content in tomato fruit', *Theor. Appl. Genet.* **73**, pp. 350–356.

Strittmatter, G. and Chua, N.-H. (1988). 'Artificial combination of two *cis*-regulatory elements generates a unique pattern of expression in transgenic plants', *Proc. Natl. Acad. Sci. USA* **84**, pp. 8986–8990.

Verma, D.P.S., Delaney, A.J. and Nguyen, T. (1987). 'A strategy towards antisense regulation of plant gene expression', in *Tailoring Genes for Crop Improvement: an Agricultural Perspective*, Eds Bruening, G., Harada, J., Kosuge, T. and Hollaender, A., pp. 155–161. New York, London, Plenum Press.

Chapter 4

Current Applications of Plant Cell and Tissue Culture

As we have seen in Chapter 2, whole plants, parts of plants, and also single cells can be grown in sterile culture, supported by media that contain minerals, growth factors, a carbon source and plant growth regulators. When plant tissues are cultured in this way, their growth can be manipulated and exploited for commercial purposes. In this chapter are described techniques that are now widely used both for research and commercial purposes. They are listed in Table 4.1 under the heading of 'applied technology', as opposed to 'developing technologies', which will be considered in subsequent chapters.

The emphasis of these techniques is placed on maintenance of genetic stability, and they are of considerable practical use to aid current breeding practices or as experimental systems in their own right. The techniques are of particular value in speeding up conventional breeding and propagation procedures, reducing space and labour requirements or achieving manipulative goals that cannot be carried out *in vivo*. These 'applied technology' aspects are considered in more detail in the following sections.

Micropropagation

A range of tissues can be explanted from different species as source material for micropropagation. Some of these are illustrated in Fig. 4.1, and they can include:

(1) Shoot meristems and stem segments with axillary buds
(2) Tissues that will form (i) adventitious shoots *and/or* (ii) adventitious embryos *either* (a) directly on explants *or* (b) indirectly via unorganized or partly organized callus.

Table 4.1 Current and future applications of plant tissue culture

Applied technology	*Developing technology*
• Micropropagation • Somatic embryogenesis • Virus elimination • Embryo rescue • Haploid production • Ploidy manipulation • Germplasm storage and transport • Production of chemicals by cultured cells	• Modification by somaclonal variation • Plant gene transfer by protoplast fusion • Gene introduction by *Agrobacterium* or other means. (Consequences include introduction of disease, pest and herbicide resistance, control of plant development)

As is discussed in detail in Chapter 5, propagation that involves an unorganized callus phase may produce variant plants. Thus, in practice, most micropropagation is achieved by maintaining organized tissues by the multiplication of meristems and axillary buds. The stages involved in micropropagation of a given plant genotype are as follows:

(1) Selection and preparation of mother plants
- choose typical, healthy, disease-free mother plants
- virus testing/elimination may be required

(2) Establish aseptic cultures
- surface sterilize and transfer explants to culture

(3) Multiplication
- multiplication of structures able to give rise to intact plants by appropriate procedures
- cultures may be recycled many times to obtain required multiplication rates

(4) Preparation for transfer
- rooting of shoots (perhaps after shoot elongation; some shoots can be transferred to the next stage without roots)

(5) Transfer to natural environment
- wash agar from roots, transfer to sterile rooting medium or 'artificial soil'
- initially maintain at high humidity
- gradually harden off

There are two main approaches to micropropagation: (a) multiplication of shoot meristems or (b) serial sub-culture of stem nodes. Somatic embryogenesis is considered separately.

PROPAGATION FROM AXILLARY BUDS

The most common method is to excise apical or lateral shoots containing meristems, and to culture these on media which suppress apical dominance to

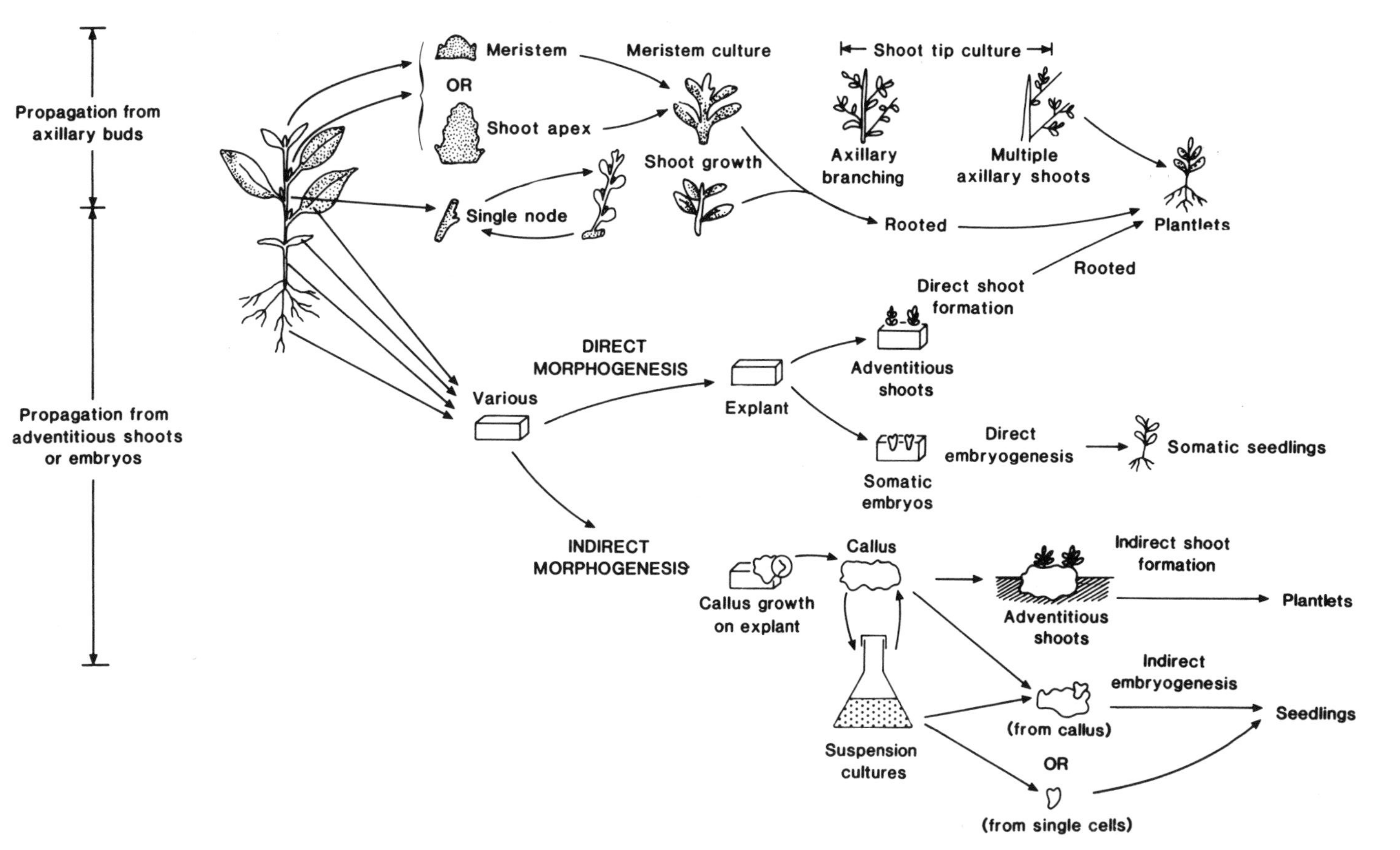

Fig. 4.1 The principal methods of micropropagation. A diagrammatic representation of different source material and methods that can be used for micropropagation. See text for further description. (Adapted from George and Sherrington, 1984, with permission.)

encourage precocious development of axillary shoots. This is usually accomplished by the addition of cytokinins to the medium, and results in the production of highly-branched shoot systems that lack roots. After 3–6 weeks of culture, individual shoots can be separated and sub-cultured on to the same medium for multiplication. This may be repeated many times, or the shoots can be transferred to rooting conditions as required.

PROPAGATION FROM NODAL SECTIONS

In some cases, the addition of cytokinin does not readily induce multiple shoots, and multiplication is achieved more easily by serial sub-culture of stem nodes. Individual stem nodes are cut, and the axillary buds allowed to grow up to several centimetres in length. Each stem is cut to separate the individual nodes, and the process is repeated. This approach is normally used, for example, for potato.

Various other explants (e.g. young inflorescences, stems, leaves) can be used, and multiplication can also be achieved by direct organogenesis or embryogenesis on such explants. Because sub-culturing is labour-intensive, the use of automated systems has potential advantages, and is the subject of much research. Alternatively the synchronous production of somatic embryos in suspension cultures can eliminate most of the labour of sub-culturing, but results in the production of naked propagules that must be coated in some way to protect them if they are to be used directly in the field as 'artificial seeds'. This aspect is dealt with separately.

Some advantages and disadvantages of micropropagation are summarized in Table 4.2.

Commercial Micropropagation

There are now many commercial companies producing millions of plantlets by micropropagation (Table 4.3). For example, in The Netherlands, where there is an efficient pot-plant and glasshouse industry, there are of the order of 10 companies producing up to 10 000 000 plants per year, and a further 40 companies serving smaller markets down to individual glasshouse complexes propagating plants only for their own requirements. Similar micropropagation companies exist in the U.K., France, Germany, Israel, the United States and elsewhere. In many cases the companies are specialized, and propagate, for example, only ferns, or specific plants for the cut flower industry (e.g. Gerberas). Work is usually carried out on a contract basis, with the production of small rooted plantlets that are passed on to other companies specializing in glasshouse growth up to marketable size.

In a commercial micropropagation laboratory it is important to carry out sub-cultures in laminar flow benches in rooms that are as clean as possible. Thus the whole laboratory may be supplied with filtered air. The micropropagators must be able to work rapidly but maintain sterility, and also be able to recognize and eliminate contaminated cultures. Culture media, perhaps developed specifically for individual plant genotypes, must be produced in media 'kitchens' with careful control of the medium components. The propagation rates must be closely

Table 4.2 Advantages and disadvantages of micropropagation

Advantages

1. *Small space requirements*. Shoot multiplication can be achieved in a small space because small shoots are multiplied
2. *Sterility*. Propagation is carried out under sterile conditions. No losses should occur from pests or diseases, and the plantlets finally produced should be free from bacteria, fungi and nematodes
3. *Certified virus free*. If virus-free material was used to initiate cultures, large numbers of virus free plants can be produced
4. *Controlled conditions*. The conditions for *in vitro* multiplication can be strictly controlled – light, medium composition, growth-regulatory levels, temperature etc. Reproducibly high rates of propagation can be achieved
5. *Continuous production*. Production can be maintained continuously without seasonal variation
6. *Labour*. No attention is required between sub-cultures (e.g. watering, weeding, spraying)
7. *Glasshouse space*. Less expensive glasshouse space is required
8. *Stock plants*. Stock plants can be stored *in vitro*
9. *Range of species*. Some plants can be propagated *in vitro* that are difficult or impossible to propagate *in vivo*
10. *Mechanization*. It may be possible to automate propagation procedures for some species and so reduce labour requirements

Disadvantages

1. *Specialized facilities*. Fairly expensive specialized facilities are required (e.g. culture rooms, laminar flow hoods etc.)
2. *Skilled operatives*. Staff must be able to work under sterile conditions and to decide where to divide different cultures
3. *Contamination*. If fungal or bacterial cantamination occurs during early multiplication many potential propagules may be lost
4. *Micropropagation conditions*. Specific methods for efficient micropropagation may need to be developed for each species, including the conditions for rooting and plantlet establishment
5. *Plantlet size*. Initial plantlets produced are small
6. *Genetic stability*. Propagation techniques must not introduce genetic instability
7. *Cost*. Because of the facilities required and labour-intensive procedures, the cost of plantlets is relatively high and must therefore be offset by a large scale of production and by high added value of the plantlets produced

monitored in order to produce the required number of plants at the appropriate time. The importance of this control is evident if, for example, 200 000 plantlets of an individual fern must be produced each week, or alternatively, if large numbers of plants must be produced in time for Easter (e.g. lilies) or Christmas (e.g. chrysanthemums or poinsettias). Since each species or even specific genotypes may have different requirements for culture media, methods of sub-division for propagation, rooting, or transfer to the environment, it is not surprising that companies specialize more in producing a limited range of plants as the numbers of individual plants required increases.

Table 4.3 Some genera micropropagated *in vitro*

Ornamentals	*Woody species*	*Vegetable and crop species*
Alstroemeria	*Araucaria*	*Actinidia*
Anigozanthos	*Betula*	*Allium*
Anthurium	*Coffea*	*Apium*
Bromeliads (many)	*Cryptomeria*	*Arachis*
Cephalotus	*Eucalyptus*	*Asparagus*
Cordyline	*Grevilea*	*Beta*
Chrysanthemum	*Hevea*	*Brassica*
Cymbidium	*Kalmia*	*Cicer*
Dianthus	*Malus*	*Cynara*
Draceana	*Musa*	*Dactylis*
Ferns (many)	*Pinus*	*Festuca*
Freesia	*Populus*	*Fragaria*
Fuchsia	*Prunus*	*Glycine*
Gerbera	*Pyrus*	*Lolium*
Gladiolus	*Ribes*	*Phaseolus*
Gloxinia	*Rosa*	*Rheum*
Hemerocallis	*Rubus*	*Solanum*
Hippeastrum	*Salix*	*Trifolium*
Hosta	*Sassafras*	*Vaccinium*
Hyacinthus	*Santalum*	*Vigna*
Iris	*Sequoia*	*Zea*
Leucojum	*Spirea*	
Narcissus	*Tectona*	
Nerine	*Thuja*	
Orchids (many)	*Vitis*	
Pelargonium		
Phlox		
Rhododendron		
Saintpaulia		
Saxifraga		
Spathiphyllum		
Syngomium		
Tulipa		

SOMATIC EMBRYOGENESIS

As seen in Chapter 2, plant regeneration through tissue culture can be accomplished by one of three routes: via culture of zygotic embryos, organogenesis or somatic embryogenesis. Somatic (or asexual) embryogenesis is the production of embryo-like structures from somatic cells. A somatic embryo is an independent bipolar structure and is not physically attached to the tissue of origin. Somatic embryos can develop and germinate to form plants in a manner analogous to germination of zygotic embryos. Production of somatic embryos from cell, tissue or organ cultures may proceed either directly (without an intervening callus phase) or indirectly after some form of callus culture. Direct embryogenesis usually occurs from an explant maintained on solid culture medium (but see also the section below on haploid production) and can be utilized for micropropagation for a more limited range of species. However, indirect embryogenesis from liquid cell suspensions is particularly attractive for micropropagation, as long as genetic stability can be maintained, because potentially vast numbers of somatic embryos can be produced in small volumes of culture media in an approximately synchronous manner (e.g. 10^5 embryos from 1 g of tissue), and this allows partial mechanization and reduced labour costs. This aspect is dealt with in more detail here because of the commercial interest and applications.

SOMATIC EMBRYOGENESIS IN CELL SUSPENSION CULTURES

Somatic embryos were first observed in cell suspension cultures of carrot (*Daucus carota*), following the scheme shown in Fig. 4.2. It soon became apparent that the phenomenon could be induced in cultures of a range of Umbellifers, and is now known to be a common event that can probably be induced in cultures of all plant families. The development of somatic embryos can be achieved by employing the appropriate culture media, explant and environmental conditions.

When grown in cell suspension cultures, cells and embryos are evenly exposed to nutrients, hormones etc. The media components can be controlled precisely to induce development of somatic embryos in a uniform manner. In liquid culture, the pro-embryo cell clusters and somatic embryos usually develop as separate structures suspended in the medium. Thus the cells can easily be sieved, centrifuged, sub-cultured and manipulated as required.

Large numbers of cells can be handled and transferred from one culture vessel to another, and various treatments can be applied which are needed for the development of embryos and the subsequent production of whole plants. In addition to providing a simple means of large-scale cloning of plants, it is possible to impose dormancy on the embryos and to create artificial seeds that can be planted by mechanized delivery systems.

Factors that are important in somatic embryogenesis include the following:

(1) *Auxin supply*. In most cases the presence of auxin (IAA or an analogue) is necessary for embryo initiation, both for the induction and for maintenance of embryogenic suspension cultures. Removal of the auxin supply, on sub-culturing, coupled with dilution of the culture density, leads to maturation of the embryos.

(2) *Nitrogen supply*. The nitrogen supplied to the culture is important. Reduced nitrogen, as ammonium salts, or in coconut water, casein hydrolysate or selected amino acids (e.g. L-glutamine, L-alanine) usually aid embryogenesis.

When cultures are transferred from maintenance medium into a medium suitable for embryo development, there is a mixture of cell cluster sizes, ranging

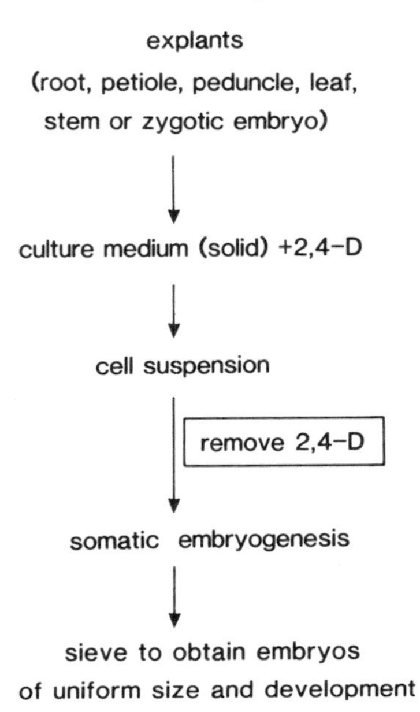

Fig. 4.2 Somatic embryogenesis from cell suspension cultures, of e.g. carrot, celery, parsley, other umbellifers, forage legumes (e.g. *Medicago*), some cereals.

from small meristematic clumps to larger pro-embryonic cell aggregates. This range in size and developmental state can be rendered more uniform in various ways. The simplest of these is to sieve the culture inoculum before sub-culture through a series of stainless-steel or nylon-mesh sieves.

Other factors to be considered include both repetitive embryogenesis, in which new embryogenic cells arise from maturing embryos, and the precocious germination of developing embryos. These phenomena lead to non-uniformity of cultures. Addition of the plant growth regulator, abscisic acid, can reduce the incidence of these phenomena and so normalize embryo development.

PLANTS FROM SOMATIC EMBRYOS

Somatic embryos can be germinated readily on culture medium without hormones until plantlets reach a suitable size for transfer to soil or vermiculite.

For bulk handling of cloned somatic embryos of agricultural species it is important to generate a population of embryos that are as synchronized, uniform and have as high a proportion of viable individuals as possible. Various mechanization treatments can then be applied. These include fluid drilling of embryos in a gelatinous nutrient medium (akin to wallpaper adhesive) or the production of 'artificial' seeds by encapsulation. Individual somatic embryos can be encapsulated in an alginate matrix by dripping embryos, coated in sodium alginate and nutrient solution, into 50 mM $CaCl_2$ solution. The calcium ions induce rapid cross-linking of the alginate, to form firm beads, of diameter about 5 mm, containing single embryos. Additives, such as fungicides and nutrients, can be included in the solution before cross-linking, as for encapsulated seeds. It is possible to dehydrate alginate-encapsulated somatic embryos, and to store them until they are required in bulk for planting in the field. Encapsulation of somatic embryos has been applied to crops such as celery, alfalfa and cauliflower. A number of biotechnology companies are actively engaged in research in this area to expand the range of plants that can be manipulated in this way and to convert the potential into a practical agricultural procedure.

VIRUS AND PATHOGEN ELIMINATION

Under normal growth conditions, plants may be susceptible to a range of pathogens including bacteria, fungi, viruses, viroids, nematodes and insects. Superficial contaminants may readily be removed by surface sterilization, but it is important that other pathogens are absent from stock plants before they are transferred to culture. It is possible to propagate plants which are diseased *in vitro*—particularly plants infected with viruses. There have also been instances when plants with nematode infections in leaves or with systemic bacterial infections have been propagated. This can spread the infections and usually depresses the multiplication rate, and it is therefore important to ensure that disease-free plants are chosen for culture. It is relatively straightforward to eliminate most pathogens by chemical treatments or other means, but viruses in particular can present a problem.

Virus diseases can be found in almost all crop plants, and depending on the severity of the infection, can cause serious losses. Systemic viruses of perennial or vegetatively propagated crops present the greatest problem, since the plants or clones derived from them will also be infected. In some cases, viruses can also be transmitted through seeds. From observations that the distribution of virus particles in plants is uneven, and especially that there are fewer virus particles in apical meristems, the tissue culture technique of meristem culture has been developed as an important and effective method for the elimination of viruses.

The apical meristems of plants are groups of cells located at the extreme growing tip, of about 0.1 mm in diameter and 0.25–0.3 mm length, and are normally dividing actively and in an organized fashion. The apical meristem of surface-sterilized shoots can be dissected out to contain the apical dome and adjacent leaf primordia. Such 'meristem tips' can be cultured on suitable media, and under appropriate conditions will grow and regenerate into whole plants (Fig. 4.3). When this procedure is carried out, it is found that viruses that were present in the original plant may be absent from the regenerated plant. However, all plants regenerated must be checked for presence of virus particles by standard ELISA techniques or by cDNA probing (see Chapter 9).

Various factors can influence the efficiency of the procedure, including the meristem tip size, its location, seasonal variation, cultural factors and heat or chemical treatments. The most important of these is explant size—generally, the

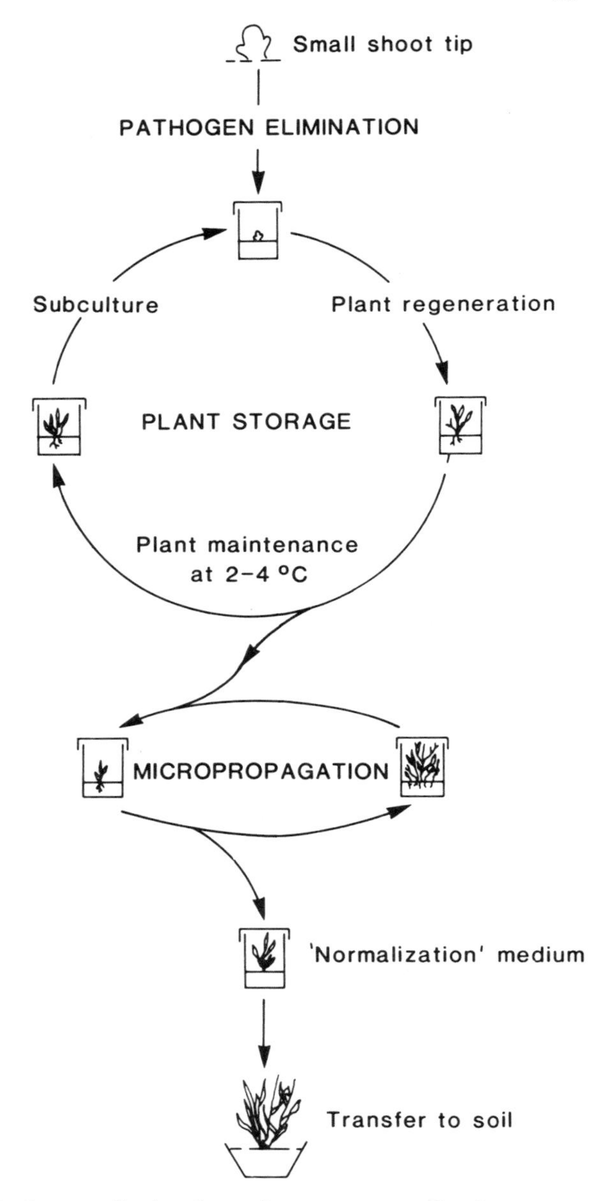

Fig. 4.3 Pathogen elimination, plant storage and micropropagation. (From Dale, P.J. and Webb, K.J. in Bright and Jones, 1985, with permission.)

regeneration frequency increases with increasing size of explant, but the proportion of virus-free plants is relatively high if the meristem tip excised is relatively small. Heat treatment (e.g. 10–20 days at 30°C) before excision is effective in some cases because many viruses are sensitive to elevated temperatures. Various chemical agents, including purine and pyrimidine analogues (e.g. 8-azaguanine, 2-thiouracil), amino acids, plant growth regulators and antibiotics (e.g. actinomycin D, glutamycin) can be applied either before excision or in culture media supplying the meristems, and in some cases such treatments aid virus eradication.

Virus-free plants have been produced in this way from at least 65 species, and, for example in potato alone, at least 136 cultivars have been freed from virus infection. Pathogen-free plants have also been produced from stocks infected with mycoplasmas, fungi and bacteria. For both crop and horticultural plants, the production of disease-free mother stocks has led to rejuvenation of older cultivars (e.g. of potato), higher yields and more rapid growth. Disease-free meristems in culture are also an ideal source material for micropropagation. This approach has therefore become an invaluable practical technique.

EMBRYO RESCUE

In sexual crosses where the parents are taxonomically distant, barriers that prevent successful hybrid production may be overcome by *in vitro* culture techniques. Specifically, in the cases where fertilization is successful, but the embryo fails to develop, immature zygotic embryos can be excised and cultured, and hybrid plants can be regenerated.

PRODUCTION OF HAPLOID PLANTS BY ANTHER AND OVARY CULTURE

Haploid plants—that is, plants with gametic chromosome numbers—are particularly useful in plant breeding, both for the rapid production of homozygous lines following chromosome doubling to the original ploidy level and for the detection and selection of recessive mutants. Although haploids occur naturally in some crop species, following parthenogenesis, semigamy, or for other reasons, *in vitro* techniques can be used for the routine production of haploids. Most *in vitro* work involves regeneration of plants from cultured anthers, or immature pollen grains (micropores), but isolated ovules can also be used (e.g. for wheat and rice). The potential advantage of microspore culture over ovule culture is that many thousands of haploid cells are available per plant and can be obtained simply by removing the anthers. In barley with, for example, the order of $2–3 \times 10^3$ pollen grains per anther, and 60 usable anthers per spike, there are potentially over 10^5microspores available for culture per spike.

Haploid plants have been regenerated in more than 50 species through anther culture, with the majority in the Gramineae, Solanaceae and Cruciferae. These include many crop species such as wheat, barley, maize, rice, rye, triticale, grasses, brassicas, potato, tomato and tobacco. As with other aspects of tissue culture, conditions and procedures vary, but the following factors are important for efficient production of haploid plants: (1) growth of donor plants under optimum conditions; (2) adequate sample of different genotypes; (3) choice of correct stage of pollen development; (4) conditions for pretreatment of the anthers, particularly cold or heat treatment; (5) the effect of additives to culture media (e.g. organic nitrogen, increased sucrose, activated charcoal); and (6) both 'floating' culture systems and agar-solidified media should be tested.

A typical regeneration procedure involves cutting off the flowerbuds (e.g. Brassicas) or spikes (e.g. barley) at the optimum stage (i.e. at the 'mid-uninucleate stage' before the microspore nucleus divides to form the generative and vegetative nuclei), and their surface sterilization (Fig. 4.4). The anthers with their developing pollen grains are enclosed by the flower structures at this stage, and should be aseptic. The buds or spikes are then usually pretreated in some way, such as 4–28 days at 4–15°C in the dark. The anthers are then excised, cultured on appropriate media with relatively high levels of sucrose or other carbon source with additives such as glutamine, casein hydrolysate, potato extract or coconut milk, growth regulators (e.g. 2,4-D) that induce embryogenesis, and perhaps activated charcoal (which adsorbs inhibitory substances). After 4–8 weeks the microspores divide and can develop either indirectly (via callus) or directly into somatic embryos. These can be germinated to produce intact plants.

Apart from the practical aspects, the sequence of events, outlined in Fig. 4.4, presents an interesting developmental system. By picking the correct developmental stage of microspore development, the pretreatment and culture medium switches development from the gametogenic pathway, i.e. production of pollen cells, to the embryogenic pathway and ultimately the production of haploid plants.

Not all the plants produced in this way are haploid, and in some cases the majority (e.g. 80%) are doubled haploids. Aneuploids can also occur, especially if there has been an intervening callus phase. Some plants may also be albino, particularly in the Gramineae, and this may be related to the programmed destruction of chloroplasts during pollen maturation.

One crop plant for which this approach works particularly well is barley. At one stage, an alternative approach to anther culture to achieve haploidy was recognized, i.e. by crossing with the wild species *Hordeum bulbosum*, after which chromosomes from *H. bulbosum* are eliminated to leave the haploid barley set. This method at first appeared to be more efficient than anther culture. However, the discovery that the replacement of sucrose as the carbon source by other disaccharides such as maltose, cellobiose or trehalose led to very efficient direct embryogenesis from mid-uninucleate stage barley microspores (as Fig. 4.5) has led to production of many doubled haploid green plants (in the order of 2000 green plants per 100 anthers cultured). The latter approach is now much more efficient than other techniques for production of doubled haploids for a range of barley genotypes. Production of doubled haploid 'pure' lines in this way from a sexual cross is much more rapid than by the more conventional approach of repeated selfing, and promises to speed up production of new barley varieties.

The rapid production of inbred lines, following chromosome doubling of

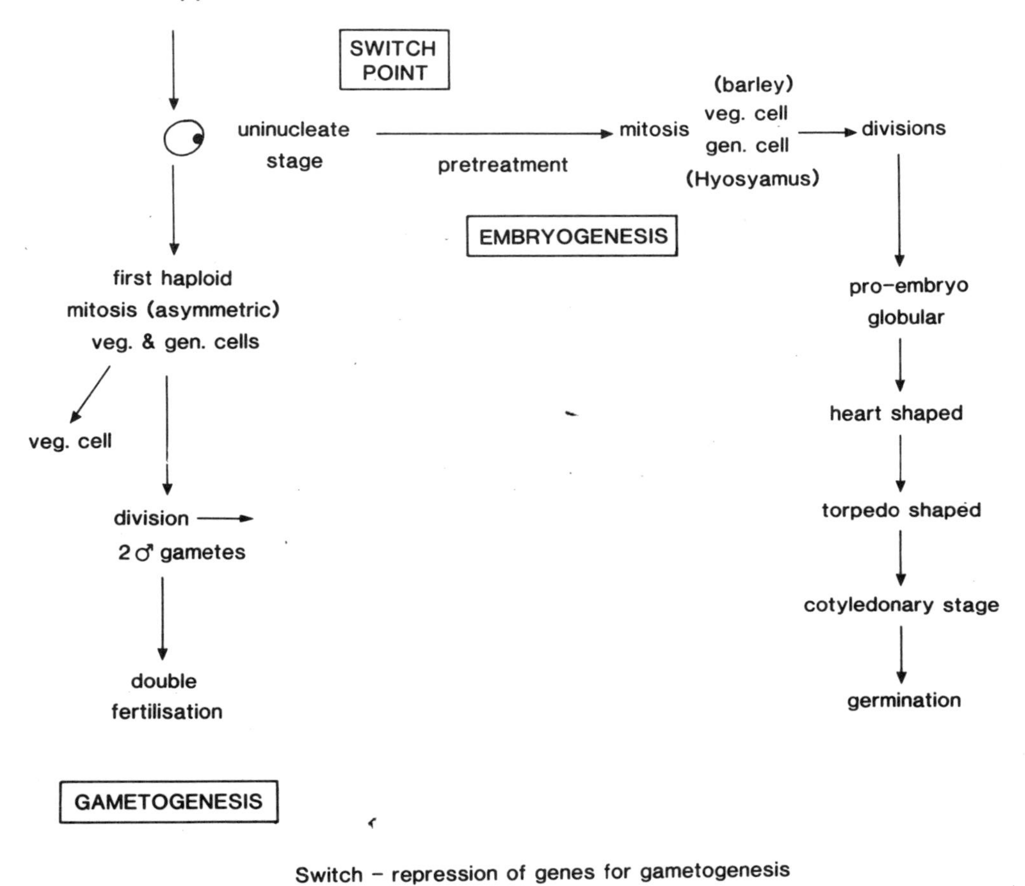

Fig. 4.4 Somatic embryogenesis from cultured microspores. The normal developmental pathway in pollen developments is 'gametogenesis'. If the developing pollen grains are cultured at the mid-uninucleate stage, their development can be switched along the new developmental pathway of 'embryogenesis', with the production of haploid plants.

haploids either using colchicine or from spontaneous doubling, has been exploited particularly by the Chinese, who have produced varieties of rice, wheat and tobacco by anther culture. The most widely grown wheat variety in China at present is a doubled haploid. Oilseed rape (*Brassica napus*), other brassicas, and various horticultural species are also well suited to this approach.

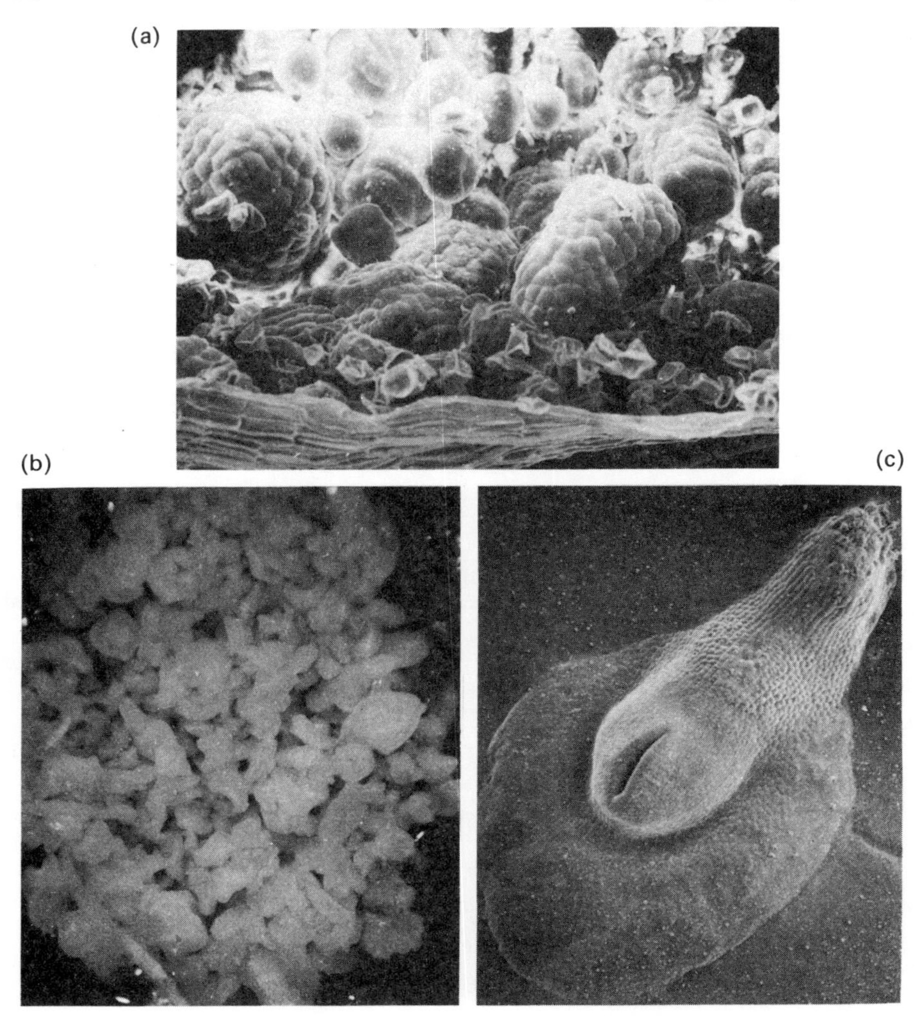

Fig. 4.5 (a) Pro-embryos derived from microspore cells of barley, still retained within the anther walls. (b) Embryoids, from (a), plated on to solid medium. A range of developmental stages is evident. (c) Microspore-derived somatic embryo of barley (scanning electron microscopy), with flat scutellum and shoot and root axes. (Courtesy Dr C. Hunter, Shell Research Ltd.)

PRODUCTION OF SECONDARY METABOLITES BY PLANT CELL CULTURES

Because of their central role in maintaining the viability of the plant, protein, carbohydrates and fats have been described as 'primary' metabolites. There are present in the plant, however, a wide range of other molecules which are synthesized via branches or offshoots of the primary metabolic pathways and the functions of which are not clear. These compounds, which include alkaloids, terpenoids, steroids, anthocyanins, anthraquinones and simple or polyphenols, are characteristically limited in their distribution. Specific compounds may accumulate only in a restricted range of species, sometimes in a single species or genus, and often within a specific organ or tissue or at a specific stage of development within a single species. Historically, such molecules have been considered not to be essential for the primary metabolic function of plants, and have therefore been described as 'secondary' metabolites. While the functions of most of these types of metabolites have not been defined, there is an increasing body of evidence to suggest that they are not simply waste products, shunted into the dumping ground of the vacuole, but may be further metabolized. Secondary metabolites are very diverse in structure, and generalizations about their role (or lack of role) would seem inappropriate. It is, however, likely that their conservation through evolution reflects a selective advantage they confer on the plant—an analogy might be the cocktail cabinet in the Rolls-Royce; while not absolutely necessary for the functioning of the motor car, it improves the quality of the journey!

The biotechnological importance of secondary metabolites is three-fold – first, many are commercially-valuable chemicals; second, a number are toxic, and should be eliminated from food products; and third, they may act as protective agents, used by the plant against pathogens and insect or animal foragers. The molecular biological aspects will be discussed separately in relation to improving product quality (Chapter 7) and to disease resistance (Chapter 9) but we will consider here how cell biological techniques have contributed to the large-scale production of specific secondary metabolites *in vitro*.

A large number of commercially-important chemicals are derived directly from plant material, and these are almost invariably secondary metabolites, often difficult to synthesize chemically. The application for these compounds can be classified into five groups: drugs, flavours, perfumes, pigments and agrochemicals, and some examples are given in Table 4.4. The possibility of using cultured cells rather than the intact plant for their production has been recognized for over 30 years, with potential advantages including synthesis under environmentally-controlled conditions, free from disease, pests, flooding and drought, and on a continuous basis, without foreign political interference (Yeoman *et al.*, 1980). The specific biosynthetic uses of cultured plant cells are: (1) biotransformation, in which one or more exogenously supplied precursors are converted, in a one- or two-step reaction, to a more valuable product; and (2) more complex, multi-enzyme synthesis, in which the product is either synthesized *de novo* (i.e. from basic

Table 4.4 Some plant secondary metabolites and their applications

Metabolite	*Application*	*Species*
	Drugs	
Atropine	muscle relaxants	*Atropa belladonna*
Hyoscyamine		*Hyoscyamus* spp.
Hyoscine		*Datura* spp.
Quinine	anti-malarial	*Cinchona* spp.
Diosgenin	contraceptive	*Dioscorea* spp.
Morphine	analgesics	
Codeine		*Papaver* spp.
Thebaine		
Digoxin	cardiatonic	*Digitalis* spp.
	Flavours	
Crocin, picrocrocin	saffron	*Crocus sativus*
Glycyrrhizin	liquorice	*Glycyrrhiza*
Capsaicin	chilli	*Capsicum frutescens*
Vanillin	vanilla	*Vanilla* spp.
	Pigments	
Anthocyanins	red/blue	various
Xanthophylls	yellow	various
Shikonin	red	*Lithospermum erythrorhizon*
Anthraquinones	red	*Morinda citrifolia*
	Perfumes	
Various complex mixtures of oils and terpenoids		various
	Agrochemicals	
Various pyrethroids	insecticides	*Chrysanthemum* spp.
Nicotine		*Nicotiana tabacum*
Piperine		*Piper nigrum*

medium components such as sucrose and salts) or from a relatively distant precursor. Biotransformations are mostly stereospecific, involving the addition or removal of single chemical groups (by hydroxylation, glycosylation, acetylation and methylation), such as the C_{12} hydroxylation of digitoxin to digoxin, a heart drug, by cultured cells of the foxglove *Digitalis lanata* (Alfermann *et al.*, 1983). Examples of more complex syntheses include the production of various alkaloids, such as scopolamine (*Hyoscyamus niger*, *Atropa belladonna*), anthocyanins (*Catharanthus roseus*, *Daucus carota*) or anthraquinones (*Morinda citrifolia*).

However, despite much early optimism, a number of biological and technical problems have been encountered in the large-scale development of cultured cell processes, not least of which is that, in the majority of cases studied so far, cultured cells have been observed to synthesize and accumulate only very low levels of specific secondary metabolites, usually levels lower than those found in the intact plant. In general terms, two factors appear to play major roles in determining the level to which a cell culture will accumulate a particular compound. The first relates to the heterogeneity and instability of gene expression and metabolic activity within a population of cultured cells. Not all cells in the population will necessarily be productive, and screening or selective procedures are commonly adopted to increase the average yield of a cell culture. A second major influence on production in cell cultures is related to the common observation that secondary metabolite production is inversely related to the rate of cell division. In a number of experimental systems, accumulation is confined to the post-division phase of the growth cycle, and in some cases, such as for tropane alkaloid biosynthesis by cultures *Datura innoxia* or capsaicin production by *Capsicum frutescens*, this has been shown to be due to a change in the direction of flux of precursor molecules common to both primary (cell-division related) and secondary (cell expansion- or maturation-related) pathways. It is also the case for some systems that multicellular organization in suspension cultures, such as the formation of roots, shoots, somatic embryos or even simple compact aggregates, favours the synthesis and accumulation of secondary metabolites, perhaps partly as a result of the limitation imposed on cell division by the multicellular condition in which the gradients of nutrients, oxygen and mechanical pressure exist. How have these observations influenced biotechnological processes to date?

There are two broad strategies for the large-scale production of plant secondary metabolites. They are based on: (1) fermenter systems, in which freely-suspended cells are grown up to a stationary phase, in a one- or two-stage process, and then harvested to extract the product; or (2) immobilized cell systems, in which cells are embedded or entrapped in an inert polymeric matrix such as a gel, foam or cartridge of hollow fibres (see Chapter 2). In the latter system the aim is to achieve a continuous or semi-continuous production process, which in turn requires that the product is naturally released or its release can be induced by reversibly permeabilizing the cells.

A number of species have been cultured successfully in large fermenters, but the product yield is often low. Nevertheless, the Mitsui Corporation of Japan have developed the commercial production of the red pigment shikonin by cultured cells of *Lithospermum erythrorhizon* in a two-phase production process: in the first phase the cells are grown up in bulk, and in the second they are transferred to a production medium which favours shikonin biosynthesis at the expense of cell division. The success of the process has been achieved largely through the selection of high-yielding lines, by exploiting the variability in production within a population of cells.

The commercial exploitation of immobilization technology, which is still relatively new for plant cells, is yet to be realized, but there are potential advantages of both a physiological and chemical engineering nature (Lindsey, 1986). For example, the aggregated nature of immobilized cells promotes

Table 4.5 Some advantageous features of immobilized plant cells for the large-scale production of secondary metabolites

- Reusability of biomass
- High biomass concentrations
- Improved metabolic performance
- Relatively long production phase compared with growth phase
- Resistance to shear damage
- Facilitated sampling of biomass
- Reduced clogging of hardware

enhanced levels of accumulation of capsaicin by *Capsicum frutescens*, anthraquinones by *Morinda citrifolia* and tropane alkaloids by *Datura innoxia*, and benefits for large-scale production include the reusability of the biomass, resistance to shear force damage (which suspended cells are particularly sensitive to), and a relatively long production phase relative to the bulking up phase (Table 4.5).

Prospects for the genetic manipulation of secondary metabolism are discussed in Chapter 7.

Storage of Germplasm *In Vitro*

The world demand for increased food production has tended to result in the production of high-yielding crop plants with a narrow genetic base. A possible consequence is a major disease outbreak or pest infestation, and there is thus good reason to preserve as much valuable germplasm as possible for future use. In addition, for micropropagation and for other plant breeding needs, it is necessary to store selected genotypes or pathogen-free plants without threat of loss that may occur if stock plants are maintained in the field. Risks associated with the latter include exposure to pathogens, pests, climatic factors and human error. The most economical form of storing germplasm for seed-propagated species is as seeds. However, some crop plants do not produce viable seeds, some seeds have a limited storage life (depending on humidity and temperature), some deteriorate rapidly because of seed-borne pathogens, and some are heterozygous and so are unsuitable for maintaining true-to-type genotypes. Various methods of *in vitro* storage of germplasm are thus of great practical interest for longer-term storage of germplasm, and good progress has been made in this area.

There are two main approaches to *in vitro* germplasm storage: slow growth techniques and cryopreservation. These will now be discussed.

SLOW GROWTH TECHNIQUES

For many years the standard method of maintaining stocks has been continuous growth of shoot cultures under optimum conditions. It is thus possible to monitor the healthy growth of stocks, but they must be sub-cultured every 4–6 weeks, and this is both labour-intensive and runs the risk of accidental loss through labelling or media errors. In order both to reduce labour input and possible errors, various methods have been developed to slow down growth in culture and so extend sub-culturing intervals. These methods include:

(1) *Storage at reduced temperature*. Cultures are maintained at 4–8°C, usually with reduced light (e.g. an 8 hour day). This may be achieved simply by modifying a suitable refrigerator, and including an alarm system in case of malfunction. This approach has been used successfully for a wide range of species including strawberry, vines, potato, grasses, legumes etc.
(2) *Culture conditions*. Vegetative growth of shoot cultures can be slowed down, whilst keeping cultures in standard temperature and light regimes by inducing osmotic stress, or altering plant growth regulators by application of growth retardants (e.g. ABA at 5–10 $mg\,l^{-1}$ or ccl at 2 $g\,l^{-1}$) or by reducing the supply of the carbon source. Perhaps the simplest of these approaches is induced osmotic stress, for example by inclusion of 4–8% mannitol in the normal culture medium. This approach has also been used widely.

By using either of these two approaches, it is possible to prolong sub-culture periods to every 12 months or longer, and the shoots can be propagated as required by transfer back to normal culture conditions.

CRYOPRESERVATION

There is only one approach that completely arrests tissue growth, and that is the reduction in temperature of biological material to that approaching liquid nitrogen. At this temperature (−196°C), almost all metabolic activities are at a standstill and can be preserved in that state almost indefinitely. However, few cultures, if any, have natural resistance to freezing, and it is therefore necessary to develop specific conditions in preparation for freezing, during the freezing and thawing processes and during recovery. These usually involve the following:

(1) *Pre-culturing*. Most cultures require special conditions to present maximum freeze-tolerance. Usually a rapid growth rate is required, which produces cells with relatively small vacuoles and therefore low water contents, to minimize damage by ice crystal formation.
(2) *Cryoprotection*. Various chemicals such as glycerol, dimethylsulphoxide (DMSO), proline, mannitol/sorbitol, sucrose, glucose and polyethyleneglycol can be added to culture media before freezing. These compounds protect against ice damage and can lower the initial freezing temperature and alter the form of ice crystals.
(3) *Freezing*. This is the most critical phase, and two methods can be followed—either rapid cooling or slow cooling. Rapid cooling results in early intracellular freezing but little dehydration, whereas with slow cooling, extracellular freezing occurs first, followed by cytoplasmic dehydration

until intracellular freezing occurs. The best approach for any particular material must be found empirically.

(4) *Storage*. Storage in liquid nitrogen (at −196°C) or its vapour (at −150°C). Changes in ice crystals will occur above −100°C, and temperatures above this value must be avoided.

(5) *Thawing*. Rapid thawing is usually adopted, to avoid the damaging effects of ice crystal growth which could occur if thawing is slow. Specimens can simply be dropped into warm water or culture medium.

(6) *Recovery*. Freshly thawed cells need to be 'nursed' back to normal growth and cryoprotectants washed out. For maximum genetic stability, callus production should be avoided.

Cryopreservation has been used successfully to store a range of tissue types, including meristems, embryos, callus and even protoplasts. However, it does depend on a reliable supply of liquid nitrogen, and is more complicated to set up than slow growth approaches. In addition, it is not yet possible to recover 100% of all frozen samples, and this is a major drawback.

There are now various germplasm collections, such as that for *Solanum* spp. at the International Potato Centre, Peru, in which *in vitro* storage is the standard method of germplasm storage. Access to these collections, and international transport of germplasm, is facilitated by *in vitro* maintenance. Stocks that have been indexed as free from pests and pathogens can meet international phyto-sanitary regulations most readily, and can be transported free from soil or any other potential source of disease.

General Reading

Conger, B.V. (Ed.) (1981). *Cloning Agricultural Plants via* in vitro *Techniques*. Boca Raton, CRC Press.

Mantell, S.H. and Smith, H. (Eds) (1983). *Plant Biotechnology*. Cambridge, Cambridge University Press.

Neumann, K.-H., Barz, W. and Reinhard, E. (1985). *Primary and Secondary Metabolism of Plant Cell Cultures*. Heidelberg, Springer-Verlag.

Specific Reading

Alfermann, A.W., Bergmann, W., Figur, C., Helmbold, U., Schwantag, D., Schulter, I. and Reinhard, E. (1983). 'Biotransformation β-methyldigitoxin to β-methyldigoxin by cell cultures of *Digitalis lanata*', in *Plant Biotechnology*, Eds Mantell, S.H. and Smith, H., pp. 67–74. Cambridge, Cambridge University Press.

Bright, S.W.J. and Jones, M.G.K. (Eds) (1985). *Cereal Tissue and Cell Culture*. Dordrecht, Nijhoff/Junk Publishers.

George, E.F. and Sherrington, P.D. (1984). *Plant Propagation by Tissue Culture*. Basingstoke, Exegetics.

Lindsey, K. (1986). 'The production of secondary metabolites by immobilized plant cells', in *Secondary Metabolism in Plant Cell Cultures*, Eds Morris, P., Scragg, A.H., Stafford, A. and Fowler, M.W., pp. 143–155. Cambridge, Cambridge University Press.

Yeoman, M.M., Miedzybrodzka, M.B., Lindsey, and McLauchlan, W.R. (1980). 'The synthetic potential of cultured plant cells', in *Plant Cell Cultures—Results and Perspectives*, Eds Sala, F., Parisi, B., Cella, R. and Cifferi, O., pp. 327–343. Amsterdam, Elsevier-North Holland.

Chapter 5

Consequences of Tissue Culture – Variability and Instability

In Chapter 4, the emphasis of the tissue culture techniques described was on the maintenance of genetic stability. This can be achieved by culture of meristems, shoots, organs or appropriate explants which retain meristematic organisation *in vitro*. However, when plant cells are cultured via some form of unorganized callus phase, either on agar media or in liquid cell suspension cultures, then plants that are subsequently regenerated may exhibit various phenotypic or biochemical characters that differ from the original parental material. The process whereby such tissue culture-induced variation is generated has been called 'somaclonal variation'. In this chapter we describe some examples of somaclonal variation, and discuss some of the underlying causes. In some cases this variation can be utilized for crop improvement, but in others it is a complicating factor that we would like to reduce or eliminate. These aspects are considered in the following sections.

Somaclonal variation

DEFINITION

Somaclonal variation is a general phenomenon of all plant regeneration systems that involve a callus phase, whether regeneration occurs through somatic embryogenesis or by adventitious shoot formation. This includes plants regenerated from protoplasts, cultured explants (immature embryos, inflorescences, leaves, stems, tubers etc.), microspores, anthers and ovaries, and also plants regenerated from cultured tumorous tissues. It is of widespread occurrence among plant species, including many important crops. It was first reported for vegetative crop plants such as sugar-cane and potato, but has now been documented for cereals such as wheat, maize, oats and rice, and dicotyledons such as tobacco, tomato, carrots, celery, legumes and ornamentals such as *Petunia* and scented *Pelargonium*. In this chapter, we refer to primary regenerants from culture as the 'R1' generation, and subsequent seed generations as R2, R3 etc.

SOME EXAMPLES OF SOMACLONAL VARIATION

One of the earliest reports of somaclonal variation was in plants of sugar-cane regenerated from somatic tissues. Amongst regenerated plants there were changes in morphology, such as auricle length and the presence or absence of hairs. Changes in chromosome numbers were also observed, as were differences in three isozyme systems (peroxidase, transaminase and amylase). Variation was also found in 'stooling' and erectness, cane diameter, stalk length and weight, and cane and sugar yield. In addition, somaclones from different varieties were tested for resistance to 'Fiji disease' and downy mildew, and clones with increased resistance to both these diseases were identified. The interesting feature of these responses is that there appeared to be a whole population shift toward improved resistance, even when the original variety was already reasonably resistant. Changes in resistance response to 'eyespot' and 'culmicolus smut' diseases have also been reported.

Somaclonal variation has also been studied extensively in potato, where variation in agronomically useful traits has been described. From a series of 'somaclones' regenerated from the variety 'Russet Burbank', screened in field trials over several years, variation was found in growth habit, tuber shape and size, tuber skin colour, photoperiod requirements and maturation date. When somaclones were assessed for disease resistance, out of 500 somaclones tested for response to early blight (caused by *Alternaria solani*) five more resistant clones were identified (Matern *et al.*, 1978; Shepard *et al.*, 1980). Similarly, out of 800 somaclones tested, 20 were found to have greater resistance to late blight (caused by *Phytophthora infestans*) race 0, and at least 4 clones were resistant to multiple races of the pathogen (Shepard *et al.*, 1980). In separate work, similar variation has been observed in somaclones, including changes in tuber colour, shape, eye depth, total yield and saleable (ware) yield and maturation data, with a population shift towards better resistance to common scab (caused by *Streptomyces scabies*).

Variation has also been reported for cereals regenerated from immature tissues (Fig. 5.1). For these, seed generation enables information to be gained concerning the heritability of changes. For wheat, extensive somaclonal variation has been described amongst regenerants of a Mexican breeding line, Yaqui 50E. This included variation in quantitative traits, including height, tiller numbers, heading date, and in qualitative traits including grain colour, α-amylase regulation, and gliadin seed storage proteins. Although many of these changes

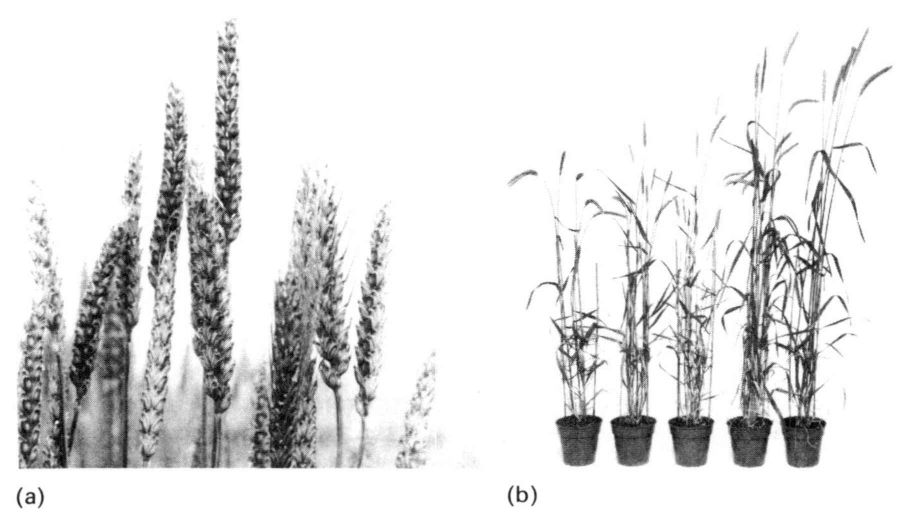

(a) (b)

Fig. 5.1 (a) Wheat plants in the field, R_2 generation, derived from a single ear of a plant regenerated from a cultured immature embryo. Somaclonal variation from awnless to awned types is evident. (b) Variation in height of triticale plants, all of which were regenerated from somatic tissues of the same cultured immature embryo.

segregated in the subsequent selfed-seed generation, and were presumed heterozygous in the regenerants, some changes were true-breeding, and therefore were presumed to be homozygous. Similar changes have been reported for rice (variation in height, tiller number, grain number per tiller, grain weight, albino variants etc.), maize and other cereals.

It is now evident that somaclonal variation can arise to some extent in most species that have been examined in detail after regeneration from a callus phase. One underlying cause of this variation is manifested in cytological changes in regenerated plants, which is considered below.

Chromosomal Variation

It has been known for some time that growth of plant cells in a callus phase can be associated with chromosome instability. Cytological studies of dividing root tip cells of regenerants from callus have shown that extensive changes to chromosomes may be evident. The extent and nature of this chromosome variation has been examined in some detail particularly in regenerants of wheat and potato, but has been described in many other species as well.

Potato is a tetraploid with 48 chromosomes ($2n=4x=48$). If plants are regenerated from pieces of leaf, stem or tuber, then about 10% of them are aneuploid, that is, they appear to have gained or lost chromosomes during the culture process. However, if potato plants are regenerated from protoplasts, the extent of aneuploidy is greater, and in some cases up to 95% of regenerants are aneuploid. However, by manipulation of culture conditions, this can be reduced routinely to about 30% aneuploid and 70% euploid (Fig. 5.2). Amongst the aneuploids, in addition to gain or loss of chromosomes at the tetraploid level (i.e. 42–50 chromosomes) much higher chromosome numbers can be found, ranging from the octaploid level ($2n=8x=96$) downwards. This may be a result of protoplast fusion or abnormal division during early stages of culture, giving cells at the octaploid level, which appear to be cytologically unstable. Normally all plants with high chromosome numbers are grossly abnormal in morphology, whereas aneuploids at the tetraploid level may be less aberrant and are sometimes indistinguishable from the parental cultivar.

A somewhat different pattern of chromosomal variation is found in regenerants of the diploid potato species *Solanum brevidens* ($2n=2x=24$). Regenerants from leaf explants of this species yield 25% diploid, 50% tetraploid and the remainder aneuploid at the tetraploid level. For both potato and *S. brevidens*, in addition to changes in chromosome numbers, structural changes such as deletions and interchanges can be found.

Similar studies on hexaploid wheat plants regenerated from cultured immature embryos showed that on average 70% of regenerants had the normal hexaploid complement ($2n=6x=42$), and structural changes were evident, including deletions, interchanges, centric fission and isochromosomes (Fig. 5.3) (Karp and Maddock, 1984).

This kind of study clearly indicates that some of the observed morphological changes are caused by changes to the chromosomes. Such changes are largely undesirable, and so any assessment of somaclonal variation should include a cytological screen. However, not all chromosome variation is undesirable, and changes in ploidy level, which can occur quite frequently during callus growth, have been exploited as a means of obtaining polyploid plants (e.g. in potato (Fig. 5.4), tobacco, *Brassica oleracea*, sugar-beet and *Petunia*). The occurrence of translocation of parts of chromosomes may also be used to obtain introgression in hybrids, such as interchanges between wheat and rye chromosomes in *Triticale*.

It should, however, be emphasized that regenerated plants (or their progeny) can be obtained that have the euploid chromosome number with no apparent structural changes. Nevertheless, when such plants are grown in the field, considerable variation can still be present. This indicates that other factors (both genetic and epigenetic) also contribute to somaclonal variation, and it is these genetic variants, with single or multiple point mutations or rearrangements, that are most likely to be agronomically useful.

USE OF GENETIC MARKERS

By using plants with specific genetic markers as parental tissues for regeneration, genetic changes can be assessed and quantified more easily. For example, plants have been regenerated from cultured cotyledon protoplasts and explants of a hybrid tobacco that was heterozygous for two marker loci involved in chloroplast

Fig. 5.2 (a) Protoplast-derived regenerants of potato (cv. 'Majestic') with different chromosome numbers: (1) $2n=46$, (2) $2n=47$, (3) $2n=48$. (b) Root-tip squashes of mitotic cells of potato root tips, following regeneration from protoplasts (*left*, normal chromosome complement of 48; *right*, variant with 93 chromosomes).

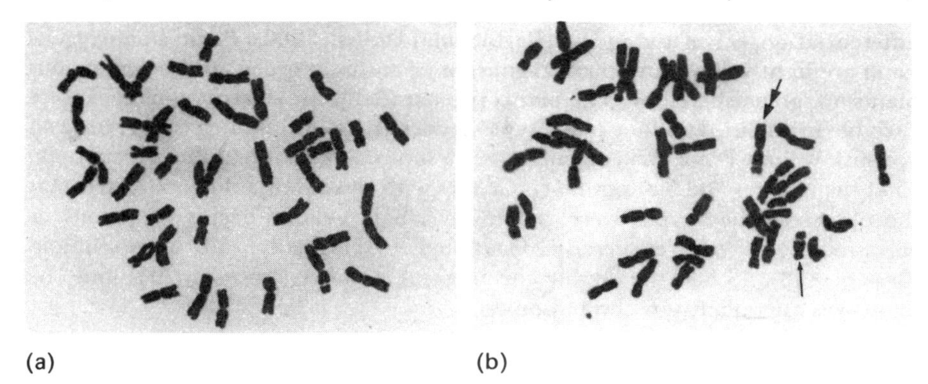

Fig. 5.3 Mitotic chromosome preparations from root tips of wheat (a) normal karyotype ($2n=6x=42$), (b) somaclonal variant with normal chromosome number, but large deletion from chromosome 16 (arrow). Compare with normal partner chromosome (double arrow). (Courtesy A. Karp.)

Fig. 5.4 Left-dihaploid potato plant ($2n=24$), right tetraploid ($2n=48$) potato plant derived from same dihaploid line after plant regeneration from a cultured leaf disc explant.

differentiation, ai^+/ai and yg^+/yg (Barbier and Dulieu, 1980). Plants homozygous for *ai* are light yellow, and homozygous for *yg* are light green, and heterozygous plants are greenish-yellow. The plants regenerated from heterozygous cells were variable in colour, and four phenotypes occurred: yellow, light green, green and greenish yellow. From 1666 regenerants, 53 showed mutation at the *ai* locus only, 55 at the *yg* locus and 5 at both loci. Crosses with the parental line indicated that the observed phenotypes were caused by single genetic events (deletions or reversions). Of 96 regenerants examined cytologically, 53 were euploid ($2n=4x=48$), 43 had the double chromosome number ($2nn=8x=96$) and one plant was aneuploid (46 chromosomes).

Factors that Affect Somaclonal Variation

It should be evident from the above discussions that the extent of chromosome stability and somaclonal variation can be affected by a number of factors. These include:

(1) The species and genotype.
(2) The ploidy level.
(3) Tissue culture procedures employed and time in culture.
(4) The source of the explant cells.
(5) The culture media used (in particular the plant growth regulators).

Some species, and indeed individual genotypes of a given species, can exhibit markedly different variation after regeneration from explants. For example, in wheat, the extent of variation recorded for regenerants of the Mexican line Yaqui 50E has been found to be far greater than that reported for regenerants from U.K. wheat cultivars. Similarly, some maize inbred lines seem more unstable in culture than others, and the extent of variation of regenerants of other species can be minimal.

There is some correlation in the extent of variation of regenerants and ploidy levels. In general, polyploids such as wheat or potato exhibit considerably more variation than plants regenerated from diploid or haploid cells. Thus barley, a diploid, gave only about 1% aneuploid regenerants when cultured under identical conditions to hexaploid wheat (10–40% aneuploids obtained). This correlated with morphological and biochemical uniformity of the regenerated barley plants. The most probable explanation is that polyploids, with two or more complete sets of chromosomes, are buffered and can thus tolerate gain or loss of chromosomes and still grow and regenerate. In contrast, loss or alteration of vital chromosomes in diploids will prevent growth, and so aneupoids are filtered out by this mechanism.

There is also good evidence that the tissue culture procedures employed can affect variation. Thus protoplast regenerants tend to be more variable than those from explants such as leaf or stem. One explanation for this is that the length of time in culture is usually greater for protoplast-derived tissues, and this is borne out by studies that show both loss of regeneration ability and increased cytological changes in callus and plants after increasing time in culture.

The explant source is also important, since there are normally cells in plants which may be polyploid. Indeed it is possible that some genetic changes may occur in differentiated cells in plants that are not readily detectable since in the plant the cells have ceased dividing. There is also good evidence that cells in cell suspension cultures are prone to cytological changes, and this is particularly marked if they are normally polyploid.

One of the most important, and complicated, factors affecting variation is the composition of the culture medium used. A sub-optimal medium can both prolong time in a callus phase before regeneration, and perhaps disturb mitosis in callus cells. In addition, plant growth regulators present in media have been shown to affect the degree of variation of regenerants as a result both of their concentration and type. High concentrations of synthetic auxins (such as 2,4-D) may lead to more variation.

The Genetics of Somaclonal Variation

When regenerated plants are potted on from cultures and grown up under standardized environmental conditions, they can exhibit transient, non-genetic or 'epigenetic' changes, as well as heritable, genetic variation. Epigenetic changes appear to be a direct effect of the culture process, being physiological in origin and perhaps resulting from effects of the growth regulators in the medium. Such changes are of no value for crop improvement as they are not expressed in sexual progeny.

To assess the value and genetic basis of somaclones, sexual progeny should be examined. Regenerated plants can be grown in greenhouses, self-fertilized, and seeds collected. These seeds can then be examined under greenhouse or field conditions. Information on transmission of variation to sexual progeny is required for further exploitation of potentially useful variants.

Genetic changes that resemble single-gene mutations have been identified in many species. If a variant trait, which was not present in the original (R_1) generation, appears on selfing, and segregates in a 3 : 1 Mendelian ratio in the seed (R_2) generation, then a recessive single gene mutation is suspected. The trait can be further tested, and should breed true when selfed and grown in further generations. This has been done for somaclones of wheat, rice, maize, tobacco and tomato. If possible, complementation tests with known mutants should be carried out, and this has been achieved for several tomato somaclones, where traits have been mapped to specific genetic loci.

Although most changes examined in this way occur originally (i.e. in the R_1 generation) in the heterozygous form, evidence from both wheat and *Brassica juncea* indicates that true-breeding changes can occur in regenerants, and these are presumed to be from direct production of homozygous variants. For *B. juncea*, 7 out of 85 regenerants were homozygous true-breeding for the recessive trait of yellow seed.

CYTOPLASMIC CHANGES

There is now good evidence that genetic changes can occur in cytoplasmic genomes (chloroplastic, mitochrondrial) as well as in nuclear genes. For example, a tobacco mutant line selected for resistance to the antibiotic streptomycin contained streptomycin-resistant chloroplasts, and genetic changes in mitochondria have similarly been recorded. The best characterized of these are somaclonal variants of maize plants with 'Texas' type male sterility (cms-T). The parental plants are susceptible to Southern Corn Leaf Blight caused by the fungal pathogen *Helminthosporium* (*Bipolaris*) *maydis* because the fungal 'T-toxin' inhibits mitochondrial function. Somaclonal variant maize plants resistant to T-toxin have been isolated among regenerants from culture. The frequency of resistant plants was very high, even in the absence of selection. All the resistant plants were also changed from male sterile to male fertile. Both toxin resistance and male fertility were cytoplasmically inherited, and isolated mitochondria from male fertile plants were resistant to toxin. These changes have been correlated with a molecular change in the mitochondrial genome in the form of a frame-shift mutation. Variations in mitochondrial DNA have also been detected among potato somaclones. More details of the molecular biological basis of cms are discussed in Chapter 8.

CURRENT METHODS OF ASSESSING GENETIC UNIFORMITY/VARIABILITY

After various *in vitro* procedures, such as micropropagation, germplasm storage or regeneration via callus, the subsequent variability of plant genotypes can be assessed in various ways. The current methods employed can be subdivided into phenotypic and genetic approaches, and are summarized as follows:

Phenotypic

- morphology
  - quantitative, e.g. height
  - qualitative, e.g. flower colour
- electrophoresis of proteins
  - denaturing
  - non-denaturing
  - iso-electric focusing
  - non-specific staining
  - enzyme activity staining
  - immunological staining
- secondary products, e.g. alkaloid production and gaseous evolution

Genetic

- chromosomes
  - general staining—aneuploidy
  - giemsa/C-banding—inversions, deletions
- restriction fragment analysis—alterations in DNA sequence

Some of these methods of assessing uniformity/variability have already been discussed in this chapter. Although morphological characters (e.g. growth habit,

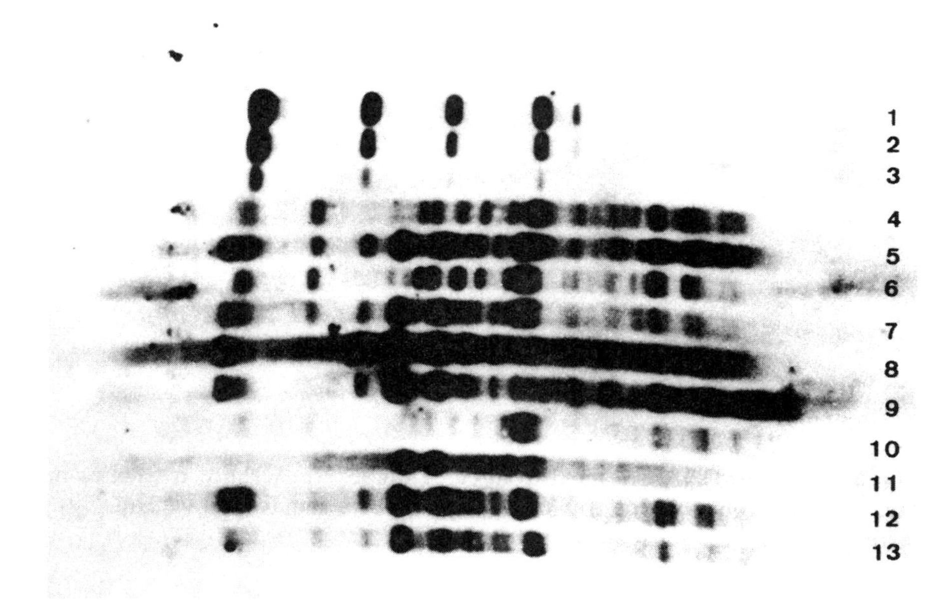

Fig. 5.5 DNA extracted from 10 potato cultivars, cut into fragments, separated by gel electrophoresis, transferred by Southern blotting to a membrane, and probed with [^{32}P]-labelled patatin gene. Each cultivar reveals a different and distinctive pattern of DNA bands. (Lanes 1–3 copy number controls; lane 4: 'Pentland Crown'; 5: 'Pentland Dell'; 6: 'Pentland Squire'; 7: 'Pentland Ivory'; 8: 'Ulster Beauty'; 9: 'King Edward'; 10: 'Majestic'; 11: 'Sarkia'; 12: 'Ulster Concord'; 13: 'Arran Victory'). (Courtesy R. Potter.)

leaf morphology, flower morphology etc.) are easy to score, there are many aspects of variation that can occur but not be manifested in obvious morphological changes. Similarly, biochemical characters, such as the presence of specific proteins as analysed by gel electrophoresis, are based on the expressed products of genes whose expression may be related to environmental or physiological factors. In addition, only a small proportion of the genes present in the genome are actually transcribed and translated at any given time. Clearly, methods of analysis of the genome at the DNA level can potentially provide much greater resolution because DNA sequences are essentially the same in all living cells of the plant, irrespective of physiological or developmental states of the tissues. There is currently much interest in developing methods for 'genetic finger-printing' of genotypes (e.g. Fig. 5.5), both for the study of stability and to look for variability. Some examples of molecular analysis at the DNA level are given below, and it should be noted that similar techniques are applied to produce genetic maps by 'restriction fragment length polymorphisms' (see Chapter 3).

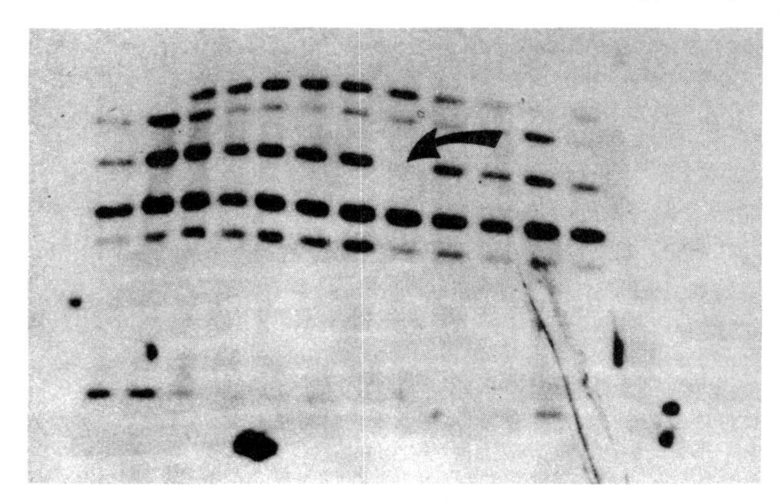

Fig. 5.6 Potato DNA extracted from cv. 'Desiree' regenerants, treated as in Fig. 5.5, but probed with a different DNA clone. A somaclonal variant (arrow) lacks one band (courtesy of R. Potter).

MOLECULAR ANALYSIS OF SOMACLONAL VARIANTS

In order to understand the basis of somaclonal variants, some somaclones have been examined at the molecular level. For example, a maize regenerant with a variant alcohol dehydrogenase enzyme was shown to be the result of a single base-pair change in the DNA sequence of the coding gene, which resulted in the change of a single amino acid in the enzyme polypeptide sequence. In addition to single-base changes, more complex alterations have been demonstrated. When DNA extracted from potato regenerants was probed by Southern hybridization with cloned potato DNA fragments, a de-amplification (i.e. reduction) in genes coding for ribosomal RNA was found, although this alteration was not associated with a change in plant morphology (Landsmann and Uhrig, 1985). Results of similar studies show that genetic changes, ranging from single base-pair changes, reduction or amplification of DNA sequences, deletions (e.g. potato somaclones, Fig. 5.6), translocations and other re-arrangements can cause the observed phenotypic and biochemical variation exhibited by some regenerants.

SOMACLONAL VARIATION AND MUTATION

Spontaneous genetic variation and mutation breeding have yielded at least 200 new plant varieties. However, spontaneous mutation occurs at a rate, on average, of 1 in 10^6, and this is generally too low for practical exploitation. On the other hand, mutation breeding is difficult to control and frequently yields mosaics and deleterious changes. Somaclonal variation can produce variants in 15% or more of the progeny, and regenerant plants are usually not mosaics. The process of regeneration sieves out genotypes with the most deleterious changes, and somaclones can usually be stabilized in a single generation. It may be asked whether somaclonal variants are the same as mutants produced by standard mutational techniques. Studies on tomato, in which variant plants were produced by mutation (ethyl methane sulphonate treatment of seeds or pollen) and by one passage through culture, indicate that there are significant differences (Gavazzi *et al.*, 1987). Various mutants recovered from both treatments were grouped into the following categories: seedling lethality, male sterility, resistance to *Verticillium*, short stature, change in the number of lateral shoots and in leaf shape. It was found that more mutations were obtained from the *in vitro* culture treatment than from chemical mutagenesis of seeds or pollen. The pattern of mutants was also different, with some mutant classes (e.g. 'potato leaf' morphology) arising exclusively by somaclonal variation. Thus a different spectrum of mutants is obtained via somaclonal variation.

Application of Somaclonal Variation to Breeding

The fact that some clonal variation can occur at a high frequency in regenerants, and that heritable single gene changes can occur in nuclear and organelle genomes, is potentially attractive to breeders. The simplest approach is to introduce the best available varieties into culture and so select for incremental improvements to them via somaclonal variation. The aim would be to retain all the favourable qualities of the variety, but to add an additional trait such as disease resistance or herbicide tolerance.

From work described so far, it is evident that there are problems as well as potential advantages of applying somaclonal variation to crop improvement, which can be summarized as follows:

Problems

- Not all characters change
- Many characters change in a negative direction
- Not all variation is novel
- Not all changes are stable
- Field evaluation is still required

Advantages

- Changes can occur in useful agronomic characters
- Changes occur at a high frequency
- Some changes are novel, and cannot be achieved by conventional technology

AN EXAMPLE OF A PRACTICAL APPROACH

A sample procedure for obtaining useful somaclones of tomato has been outlined by Evans *et al.* (1984):

(1) Regenerate plants (R_1 generation) from leaf discs, transfer them to soil and grow them to maturity.
(2) Self-fertilize plants to obtain R_2 seed. Harvest the fruit and sow the seeds after hot-water treatment to break dormancy.

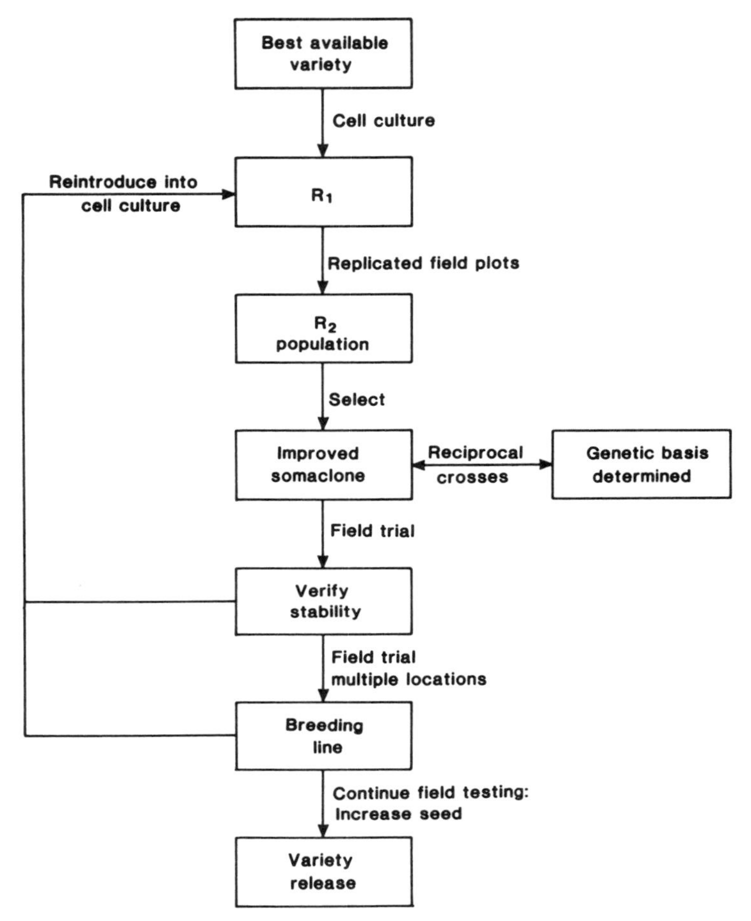

Fig. 5.7 Strategy for applying somaclonal variation to plant breeding (modified from Evans *et al.*, 1984, with permission).

(3) Record the morphology of R_2 seedlings in the greenhouse, and then transplant in replicated plots in the field.
(4) Record the morphology and fruit data, and collect seeds from the plots with interesting variants.
(5) Single plant selections of R_2 plants are evaluated as R_3 seedlings and mature plants to determine the inheritance pattern of interesting traits.
(6) Cross plants with interesting traits with known tomato mutants to determine whether previously known and new variants are allelic.
(7) If the known mutant and the variant do not complement, then they are allelic, and the new variant can be located to a specific map position. If the mutant and variant complement, they are non-allelic, and are therefore controlled by separate genes. The variant can be crossed to other tester lines to map the new locus.

(8) Traits that appear to be cytoplasmic variants can be assessed for maternal inheritance by reciprocal crossing between the new variant and the parent lines.
(9) Plants selected for useful traits are self-fertilized and evaluated in field trials until a pure line is obtained.
(10) Seeds of selected lines are bulked up for release of a new variety.

This scheme is shown diagrammatically in Fig. 5.7, but it should be noted that the types of genetic changes recovered depend on the starting material. If a diploid inbred variety of tomato is used, most regenerated plants are heterozygous and new mutants segregate in the R_2 generation. If the parental cells are heterozygous, mutations may be found in the R_1 generation. However, if haploid cells are used as source tissues, then the variation has been described as 'gametoclonal'. In this case, both recessive and dominant mutations will be expressed directly in the R_1 generation. To use 'gametoclones', the chromosomes must be doubled. This frequently occurs spontaneously (up to 80% of barley microspore regenerants may be doubled haploids) or doubling can be achieved by treatment with colchicine. A breeding strategy for gametoclonal variation is shown in Fig. 5.8. In this case, most of the variation recovered results from meiotic recombination during sexual reproduction of the F_1 hybrid if regeneration occurs by direct embryogenesis.

SELECTION IN CULTURE

So far we have considered mainly variation that occurs spontaneously through culture. There is evidence for 'hot spots', i.e. some types of changes occur at a relatively high frequency, whereas others do not appear to vary or do so at a low frequency. It is possible to include selective agents in culture media to try to select desired mutants that may be induced by physical or chemical mutagens or by exploiting somaclonal variation. The potential advantage of selection *in vitro* is that millions of individual cells can be treated and selected under controlled conditions in a relatively small space. The main problem is, however, that a good understanding of the biochemical basis of the desired mutant is required in order to design the best selection system, but at present this is not available for most agronomically desirable characters. Successful selection can be achieved for specific traits such as resistance to herbicides, toxins, heavy metal ions and amino acid analogues. Thus maize and alfalfa plants resistant to herbicide have been selected in this way, as have tomato variants resistant to tobacco mosaic virus and lettuce plants tolerant of high levels of aluminium. But if a less specific selection pressure (e.g. salt) is applied, physiological changes resulting in resistance in culture may not be reflected in regenerated plants.

SOMACLONES AND THE MARKETPLACE

Although potentially useful somaclonal variants have been produced for many crop species, as with any conventionally bred variety, it takes five years or more to assess interesting genotypes in different localities and growth conditions. The

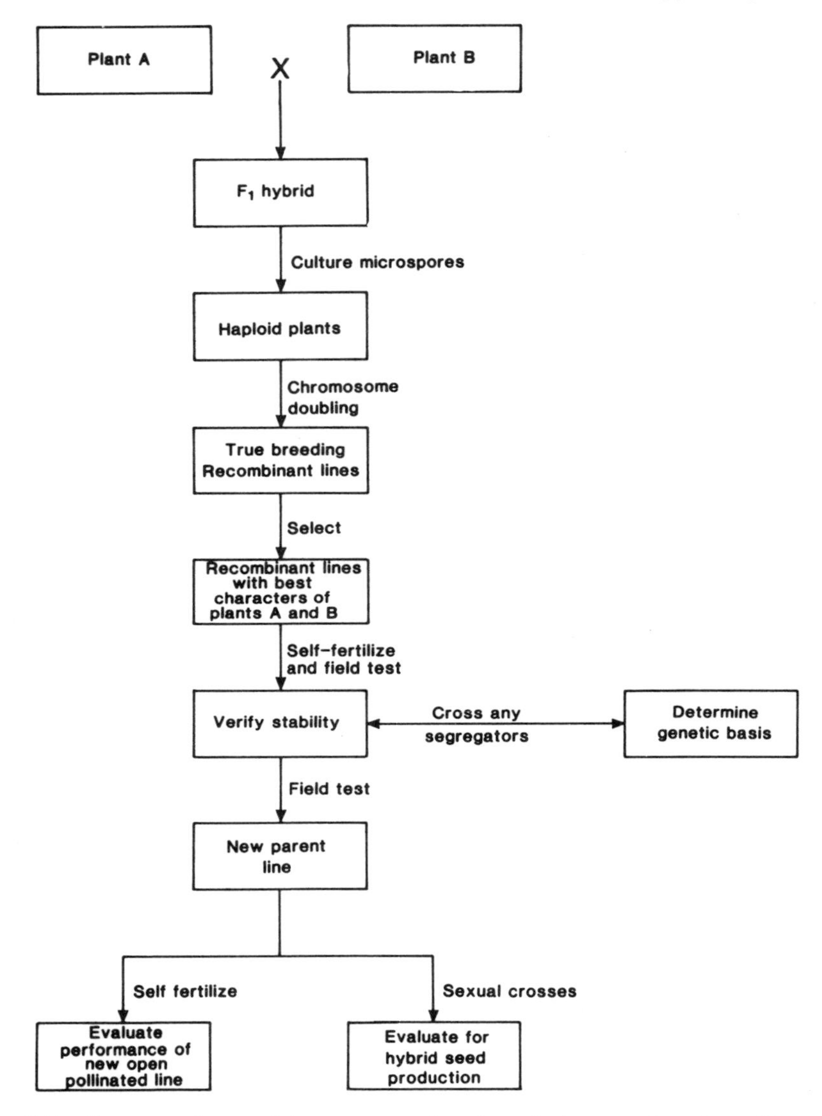

Fig. 5.8 Strategy for applying gametoclonal variation to plant breeding (modified from Evans *et al.*, 1984, with permission).

prospect of 'designer vegetables' produced by upgrading existing varieties is attractive, and there are some products already on the market. These include ultra-crisp celery and sweeter carrots, produced by DNA Plant Technology in the USA and marketed as 'Vegi-Snax', and many improved variants are currently being evaluated, including tomato (higher solids content, altered pigments and other characters), potato (disease resistant, improved cooking quality), and variants of strawberry, lettuce and maize.

It is probable that during the next few years somaclonal variation will be used widely for crop improvement, because the techniques are simple and plants can be transferred directly to the field and evaluated as part of an ongoing breeding programme.

General Reading

Evans, D.A. and Sharp, W.R. (1986). 'Applications of somaclonal variation', *Bio/Technol.* **4**, pp. 528–531.

Jones, M.G.K. and Karp, A. (1985). 'Plant tissue culture and crop improvement', *Advances in Biotechnological Processes*, Vol. 5, pp. 91–121. New York, Alan R. Liss, Inc.

Karp, A., Bright, S.W.J. (1985). 'On the causes and origins of somaclonal variation'. *Oxf. Surv. Plant Mol. Cell Biol.* **2**, pp. 199–234.

Semal, J. (Ed.) (1986). *Somaclonal Variations and Crop Improvement.* Dordrecht, Nijhoff.

Specific Reading

Barbier, M. and Dulieu, H. (1980). 'Genetic changes observed in tobacco plants regenerated from cotyledons by *in vitro* culture', *Ann. Amelior. Plant*, **30**, pp. 321–344.

Evans, D.A., Sharp, W.R. and Medina-Filho, H.P. (1984). 'Somaclonal and gametoclonal variation', *Amer. J. Bot.* **71**, pp. 759–774.

Gavazzi, G., Tonelli, C., Todesco, G., Arreghini, E., Raffaldi, F., Vecchio, F., Barbuzzi, G., Biasini, M.G. and Sala, F. (1987). 'Somaclonal variation versus chemically induced mutagenesis in tomato (*Lycopersicon esculentum* L.)', *Theor. Appl. Genet.* **74**, pp. 733–738.

Karp, A. and Maddock, S.E. (1984). 'Chromosome variation in wheat plants regenerated from cultured immature embryos'. *Theor. Appl. Genet.* **67**, pp. 249–255.

Landsmann, J., Uhrig, H. (1985). 'Somaclonal variation in *Solanum tuberosum* detected at the molecular level', *Theor. Appl. Genet.* **71**, 500–505.

Matern, U., Strobel, G. and Shepard, J.F. (1978). 'Reaction to phytotoxin in a potato population derived from mesophyll protoplasts', *Proc. Natl. Acad. Sci. USA* **75**, pp. 4935–4939.

Shepard, J.F., Bidney, D.F. and Shahin, E. (1980). 'Potato protoplasts in crop improvement', *Science* **208** pp. 17–24.

Chapter 6

The Cell Biology of Genetic Engineering

In addition to culture-induced genetic changes, protoplasts and tissue culture systems present a variety of opportunities for genetic manipulation, and are a necessary part of the overall processes required for the genetic engineering of plants. In this chapter we consider in more detail the applications of protoplasts both as a means of gene combination and gene transfer, and methods used to introduce foreign DNA into plant cells with the subsequent recovery of transformed or 'transgenic' plants.

Isolation, Culture and Plant Regeneration from Protoplasts

Standard approaches used to isolate and culture protoplasts have been outlined in Chapter 2. In this section two aspects are expanded – a consideration of events involved in protoplast isolation and culture that should be borne in mind when experimenting with protoplasts, and the range of agronomically important plants that can be manipulated by protoplast technologies.

WHAT HAPPENS TO PROTOPLASTS DURING ISOLATION AND CULTURE?

Some of the events that occur when protoplasts are isolated and cultured are indicated in Fig. 6.1. Protoplasts can be isolated from a range of tissues, which may be in one of a range of differentiated or metabolic states at the time of explanation. Storage tissues, such as tubers or cotyledons, may be virtually

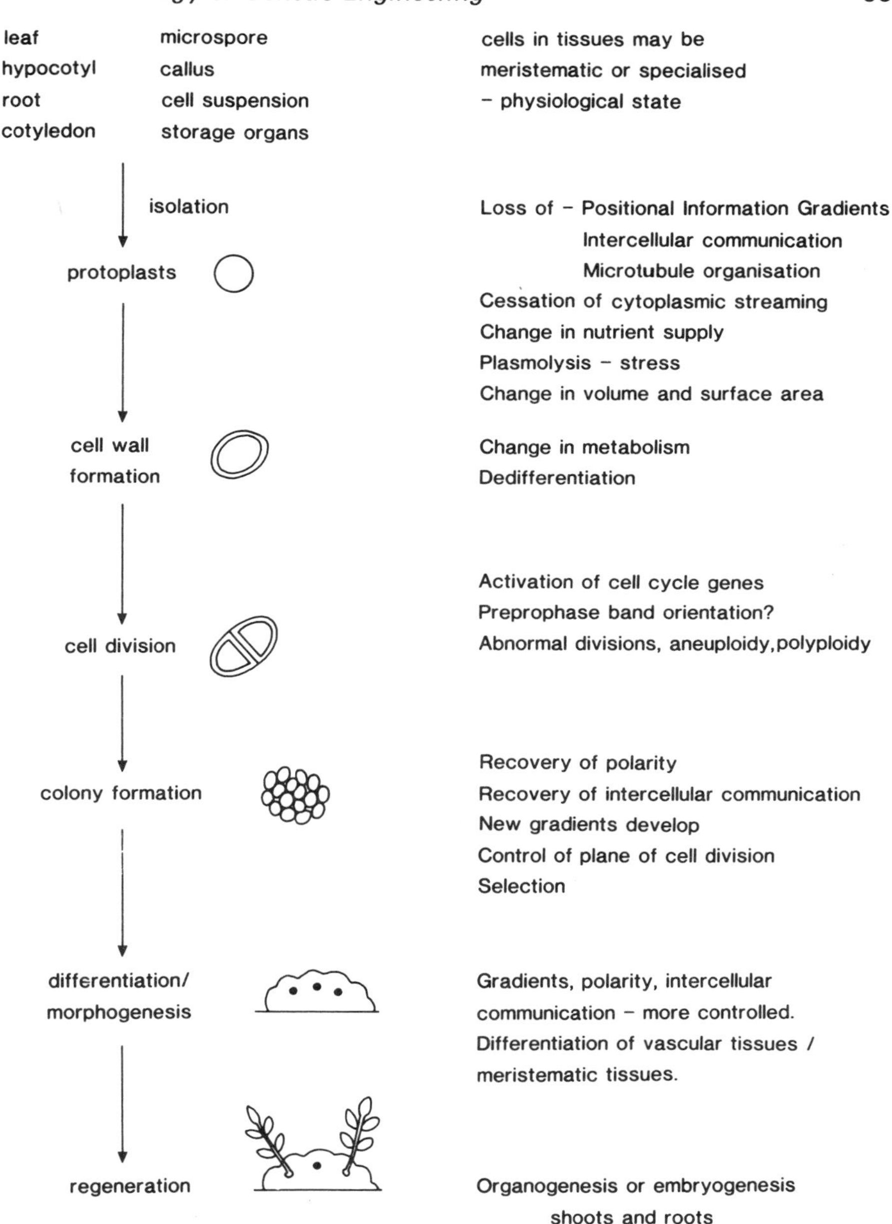

Fig. 6.1 Some of the events that occur when protoplasts are isolated and cultured.

dormant; leaf mesophyll cells are specialized to carry out reactions associated with photosynthesis, whereas cells of suspension cultures are actively dividing, synthesizing new cell walls and growing. The physiological and metabolic state of the starting tissues therefore affects subsequent performances of protoplasts in culture.

The process of plasmolysis and isolation of protoplasts is stressful to the cell, involving loss of positional information, breakdown of intercellular communication via plasmodesmata, alteration of the cytoskeleton pattern, a change in nutrient supply and in volume and surface area. Cytoplasmic streaming frequently ceases for up to 24 hours on isolation. With the presence of different plant growth regulators and nutrients in the culture medium, metabolic changes occur in protoplasts such that they may de-differentiate. One of the first signs of this is synthesis of a new cell wall, so that the protoplast is no longer spherical, and this may be followed by cell division under appropriate conditions. Initial divisions may be asymmetric or abnormal, perhaps because positional information, as manifested normally by the presence of a 'pre-prophase band' of microtubules that mark out the site of a new cell wall plane, is absent. With continued division and colony formation, new gradients are established, intercellular communication recovers as plasmodesmata are formed in new division walls, and division planes are more organized. The internal environment of colonies becomes modified by surrounding cells and can result in meristematic areas and differentiation of cell types such as vascular tissues. With continued morphogenesis, either organs or somatic embryos can develop, perhaps with the production of complete protoplast-derived plants.

CROP PLANTS AMENABLE TO PROTOPLAST CULTURE

Although it is probably possible to isolate protoplasts from all plant species, and new examples are constantly being reported, this does not necessarily mean that plants can be regenerated from such protoplasts. In some cases, reports of plant regeneration from protoplasts serve to demonstrate that this can be achieved in principle, rather than that an efficient protoplast-to-plant culture system is available. The aim with protoplast manipulations is to achieve this objective for all genotypes of plants of interest, but reproducible success is the exception rather than the rule. However, efficient plant regeneration can be achieved for many genotypes of various crops and some of these are listed in Table 6.1.

It is evident from Table 6.1 that many crops within the Solanaceae, such as tobacco, tomato and potato, can be regenerated back to plants. Similarly, various brassicas, in the Cruciferae, are amenable, as are many forage legumes and members of the Umbelliferae. Two exceptions that involve important crops should also be evident. These are the major cereal crops in the Gramineae, and grain legumes. With the exception of rice, which has now been regenerated from protoplast to plant in many laboratories, and isolated reports for wheat, sugar-cane and maize, the related grain crops, barley, rye, triticale and sorghum, have not been regenerated to plants. Plant regeneration from cereal protoplasts is currently being studied intensively, particularly in connection with cereal transformation

Table 6.1 Species of economic interest for which protoplast to plant regeneration has been achieved

Apium graveolens (celery)	*Lycopersicon esculentum* (tomato)
Asparagus officinalis (asparagus)	*Medicago sativa* (lucerne or alfalfa)
Beta vulgaris (sugar-beet)	*Nemesia strumosa*
Brassica napus (oilseed rape)	*Nicotiana* spp. (many species of tobacco)
B. campestris (turnip)	*Oryza sativa* (rice)
B. olearacea (cabbage/sprouts/kale)	*Panicum maximum* (Guinea grass)
Bromus inermis (brome grass)	*Pennesitum americanum* (pearl millett)
Capsicum annuum (pepper)	*P. purpureum*
Citrus sinensis (citrus)	*Petunia* spp.
Cichorium endiva (endive)	*Saccharum officinarum* (sugar-cane)
C. intybus	*Santalum album*
Daucus carota (carrot)	*Senecio vulgaris* (groundsel)
Digitalis lanata (foxglove)	*Solanum melongena* (egg plant or aubergine)
D. purpurea	*S. tuberosum* (potato; and related *Solanum* spp.)
Helianthus annus (sunflower)	*Trifolium pratense* (red clover; and related species)
Hemerocallis spp.	*Trifolium repens* (white clover; and related species)
Ipomoea batatas (sweet potato)	*Triticum aestivum* (wheat)
Lolium multiflorum (Italian ryegrass)	*Zea mays* (maize or corn)
L. perenne (ryegrass)	
Lotus corniculatus (birdsfoot trefoil)	

(see later), and limited success—e.g. albino barley plants, vestigial green meristems and leaves from wheat, sterile plants of maize—has been achieved. These results are encouraging, but there is still a need to obtain more routine regeneration of fertile plants from such protoplasts.

Similarly, despite considerable success in regeneration of plants from protoplasts of forage legumes (alfalfa, clovers, trefoils, etc.), the grain legumes (soybean, peas, beans, lupins) have proved so far to be intractable.

Potential applications of protoplast technology for genetic manipulation are considered below.

Protoplast Fusion

Removal of cell walls renders protoplasts amenable to various techniques of genetic manipulation, particularly those based on fusing protoplasts of different types together. These include:

(1) Combination of two complete genomes.
(2) Partial genome transfer from a 'donor' to a 'recipient' protoplast, to produce partial 'asymmetric' hybrids.
(3) Transfer of organelles ('cybrid' production)—notably chloroplasts and mitochondria—in relation to transfer of properties such as herbicide resistance and cytoplasmic male sterility.

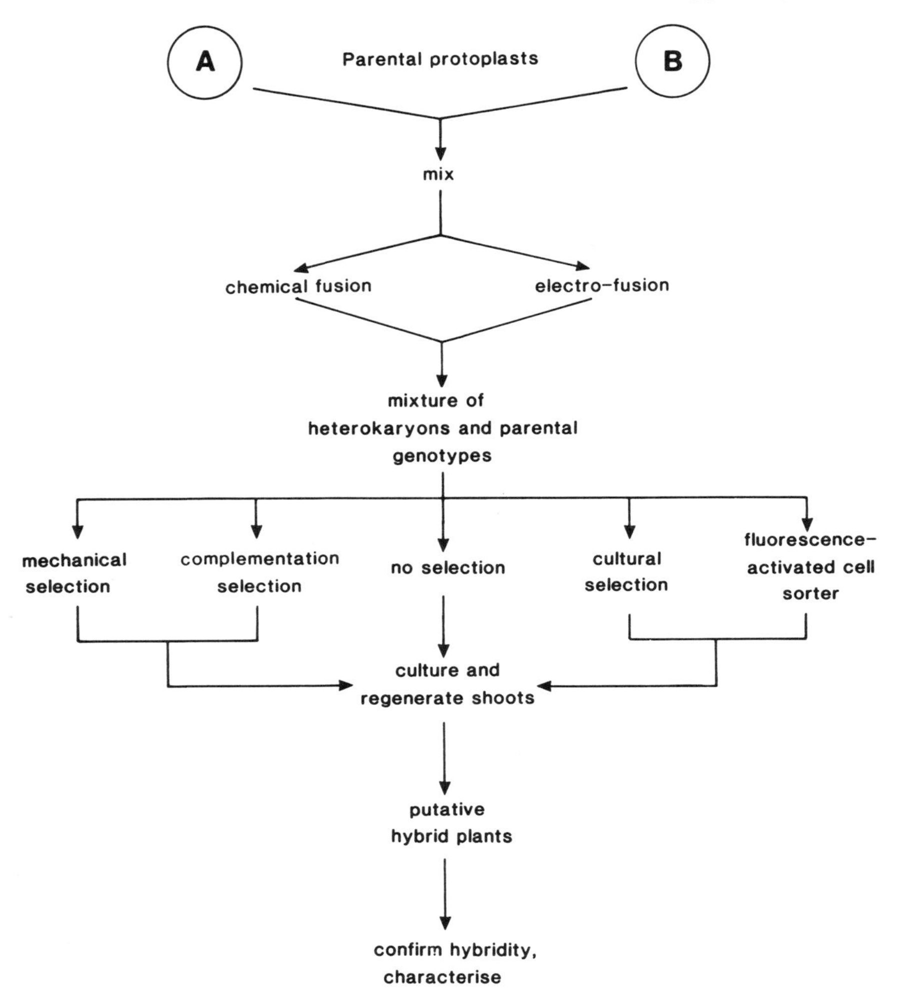

Fig. 6.2 General scheme followed for the production of somatic hybrids between different protoplast populations, A and B.

The general approach is outlined in Fig. 6.2. When protoplast fusion has been achieved, a cell with two (or more) nuclei is produced. If the protoplasts are derived from different parental cell types they are called 'heterokaryons', and if they originate from genetically identical cells they are called 'homokaryons'. The term 'hybrid' applies once fusion of nuclei in a heterokaryon has occurred.

A different type of fusion can also be achieved when the cytoplasmic organelles, mitochondria and chloroplasts, which exhibit partially independent expression of their own genetic information, are transferred without nuclear transfer. Cells produced with the nucleus of one protoplast and organelles of another protoplast

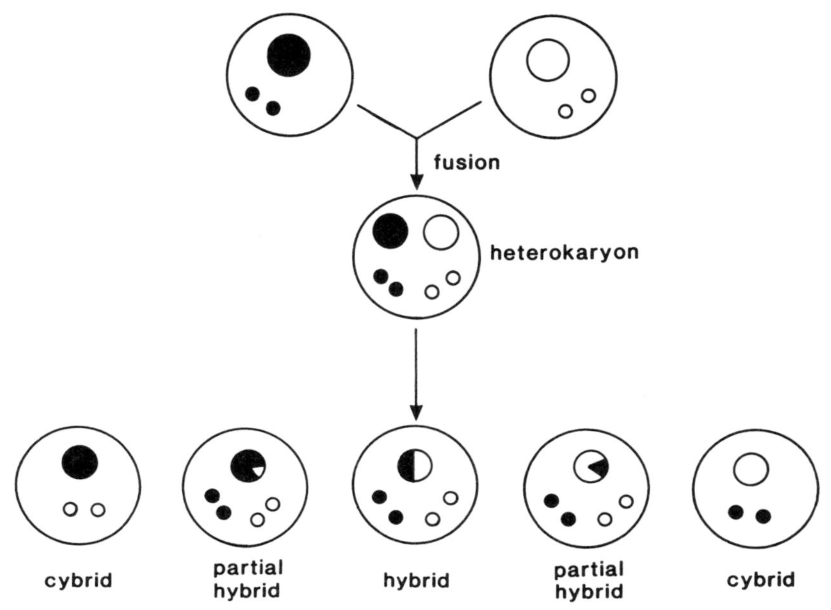

Fig. 6.3 Fusion of two genetically-different protoplasts can result in complete hybrids, partial hybrids and cybrids.

type are called 'cybrids', to distinguish them from nuclear hybrids. These different combinations of genetic information are shown diagrammatically in Fig. 6.3.

Partial hybrids can also occur if some of the chromosomes from one parental nucleus are eliminated. During subsequent culture of heterokaryons, recombination between microchondrial genomes may well occur, although chloroplast genomes usually do not recombine, and segregation occurs so that regenerated plants may contain hybrid nuclei, hybrid mitochondrial genomes, but chloroplasts from one or other parent only.

In the following sections we consider first of all the methods used to fuse protoplasts together, and then some examples of the various approaches. Two different approaches to achieve protoplast fusion are available – chemical and electrical. Chemical methods of fusion were developed first, in the early 1970s. More recently, the technique of electrofusion has been applied with great success.

CHEMICALLY-INDUCED FUSION

To initiate fusion of protoplasts, the plasma membranes must be in close contact. The net negative surface charge on protoplasts tends to repel adjacent protoplasts. Various approaches can be used to overcome this repulsion, and to destabilize the adjacent plasma membranes and so to induce fusion. Some chemical fusogens that have been used are listed in Table 6.2, and include reagents that modify surface

Table 6.2 Chemical fusogens

Chemical fusogens
Salt solutions e.g. sodium nitrate, potassium nitrate
High pH and Ca^{2+}
Polyethyleneglycol (PEG)
PEG then high pH and Ca^{2+}
PEG plus DMSO
Polyvinyl alcohol
Dextran sulphate
Polycations, e.g. poly-L-lysine

charges by manipulation of pH, addition of polycations or by dehydration of protoplasts.

The most widely used method of chemical fusion involves the sequential addition of polyethylene glycol (molecular weight = 1500–6000) to protoplasts, which causes extensive agglutination and distortion, followed by its elution with buffer at high pH (9–11) containing relatively high levels of Ca^{2+} ions (10–50 mM). This results in fusion of 1–10% of the protoplasts. The mechanism by which fusion occurs may involve lateral diffusion of proteins in contacting membranes, to create lipid-rich domains which are destabilized and fused.

Chemical fusion procedures are not all equally applicable—for example, some protoplasts will not tolerate exposure to polyethylene glycol. Nevertheless, the approach has been used to produce a range of somatic hybrid plants. (Approaches used to identify and characterize somatic hybrids are considered later.) Because chemical fusion frequencies may vary for a given species, and are relatively low particularly when applied to crop genotypes selected for agronomic properties rather than good tissue culture responses, electrofusion has recently been examined in more detail.

ELECTROFUSION

The first published report that electric fields could be applied to induce fusion of plant protoplasts was by Senda *et al.* in 1979. Zimmermann and co-workers in West Germany then developed electrofusion methodologies for both animal cells and plant protoplasts. Electrofusion involves two steps. In the first, protoplasts are placed in a medium of low conductivity between two electrodes, and a high-frequency alternating field (0.5–1.5 MHz) is applied across them (Fig. 6.4). By a process known as 'dielectrophoresis', the surface charge on the protoplasts becomes polarized and they act as dipoles, migrating to regions of higher field intensity. During the migration of protoplasts along the field lines they contact other protoplasts to form chains ('pearl chains') parallel to the applied field lines. The length of the chains depends on various factors, including protoplast density, field strength and the length of time the field is applied.

When animal cells are electrically fused, asymmetric fields are usually applied between closely-spaced electrodes (e.g. 100–200 μm separation). In this case the

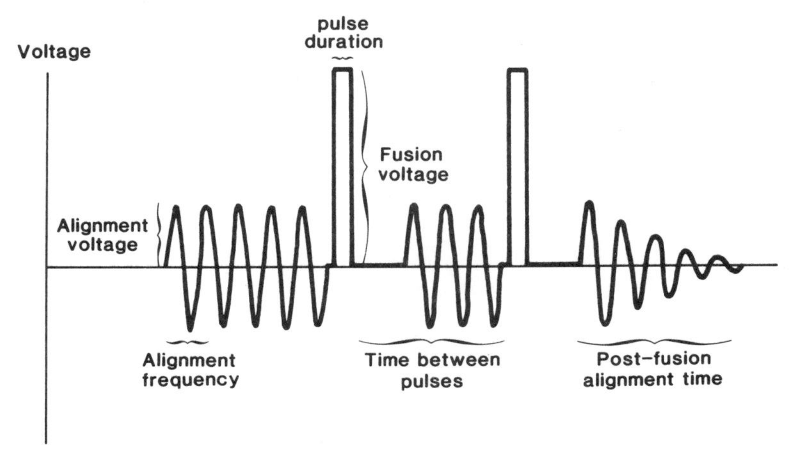

Fig. 6.4 Profile of the electric fields applied to protoplasts in electrofusion. The alternating current 'collecting field', about 1 MHz, is applied to align the protoplasts by dielectrophoresis. Once aligned, fusion is induced by applying one or more rectangular direct current pulses (1–3 kV cm^{-1}, 10–100 μs), then the alternating field is reapplied briefly to maintain close membrane contact for fusion.

cells migrate towards, and form chains on, the electrodes. However, this electrode arrangement has been found to be less useful for fusing plant protoplasts, and more uniform fields between widely spaced electrodes (up to 5 mm) are more commonly employed. With such an electrode arrangement the majority of protoplasts form chains without migrating to the electrodes. The numbers of protoplasts can then be increased greatly and their handling and observation is much easier. A consequence of the wider electrode spacing is that the a.c. field inhomogeneities that result in dielectrophoretic movement of protoplasts are caused by the protoplasts themselves. Thus individual protoplasts will migrate to one of the electrodes in the inhomogeneous close electrode system, whereas the protoplast density must be reasonably high in the wide electrode system before chain formation occurs.

The second step of electrofusion, when protoplasts have been aligned in chains, is to apply one or more short (10–200 μs) direct current pulses (1–3 kV cm^{-1}; Fig. 6.4) sufficient to cause reversible membrane breakdown ('pore formation'). Contacting membranes may then fuse. The a.c. field is briefly reapplied to maintain close protoplast contact as fusion begins, and is then reduced to zero. Rectangular fusion pulses are most widely used, but it is possible to fuse protoplasts using capacitor discharge 'spike' pulses.

It is also possible to electrofuse protoplasts without first applying an a.c. collecting field if the fusion pulse is applied to layers of protoplasts in close contact. However, this results in some loss of control over fusion events, although it allows the fusion pulse to be applied in full culture medium.

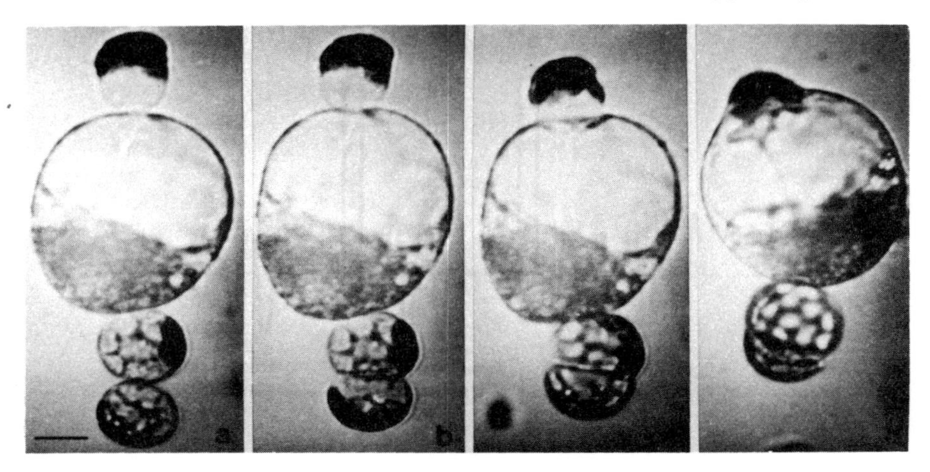

Fig. 6.5 Fusion of large hypocotyl protoplast of *Brassica napus* with small mesophyll protoplast of *Solanum brevidens*, and self-fusion of two *S. brevidens* mesophyll protoplasts. Fusion in 0.5 M mannitol over about 15 min. (From Tempelaar and Jones, 1985, with permission.)

Electrofusion Characteristics of Plant Protoplasts

One advantage of electrofusion over chemical methods is the greater degree of control that can be maintained over the fusion process. Using a simple analytical cell, with electrodes consisting of two wires stretched 1-mm apart across a microscope slide, it is straightforward to analyse the electrofusion parameters of plant protoplasts. The a.c. alignment field can be applied, followed by the fusion pulse(s), then the a.c. field can be reduced to about 30% of the alignment value, which holds the protoplasts in chains and allows fusion to be quantified (Fig. 6.5).

Using this approach, 'pulse duration-fusion response' curves (Fig. 6.6a) can be generated for different pulse voltages, electrofusion media and other parameters. Only protoplasts aligned in chains are scored, and since the fusion process takes about 10 min, the numbers of protoplasts fusing (i.e. unfused, 2, 3, 4 etc. protoplasts fused together) can be counted. With chains of 5–10 protoplasts being subjected to the fusion pulse, it might be expected that more multifusions might result. In practice, by choosing the correct pulse parameters, 50–60% of the fusion products can be 1 : 1 fusions at relatively high overall fusion frequencies.

As indicated in Fig. 6.6a, different populations of protoplasts exhibit different degrees of fusability. In general, leaf protoplasts, such as those of the wild potato species *Solanum brevidens*, fuse more readily than those from cell cultures, even when protoplasts are obtained from the same genotype. Protoplast size is another influencing factor. In general, for the same protoplast types, larger protoplasts fuse more readily than smaller protoplasts—the diameter of the suspension cultured protoplasts indicated in Fig. 6.6 were 44 μm for *Datura innoxia* and 23 μm for *S. brevidens*. Fusion frequencies are also affected by:

(a)

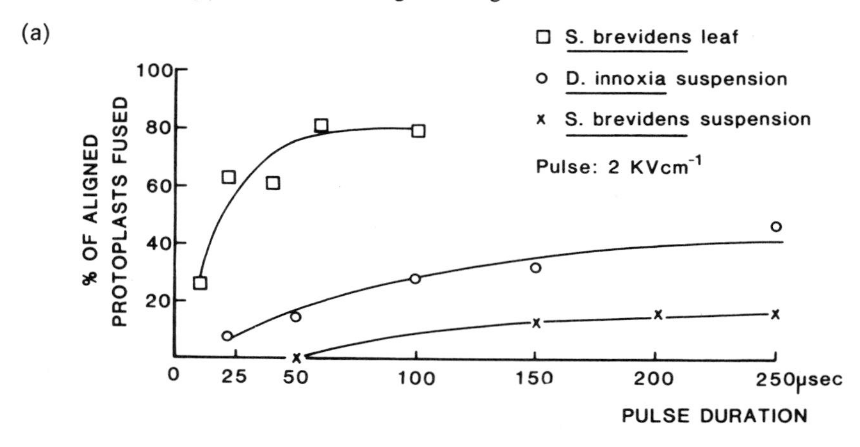

(b)

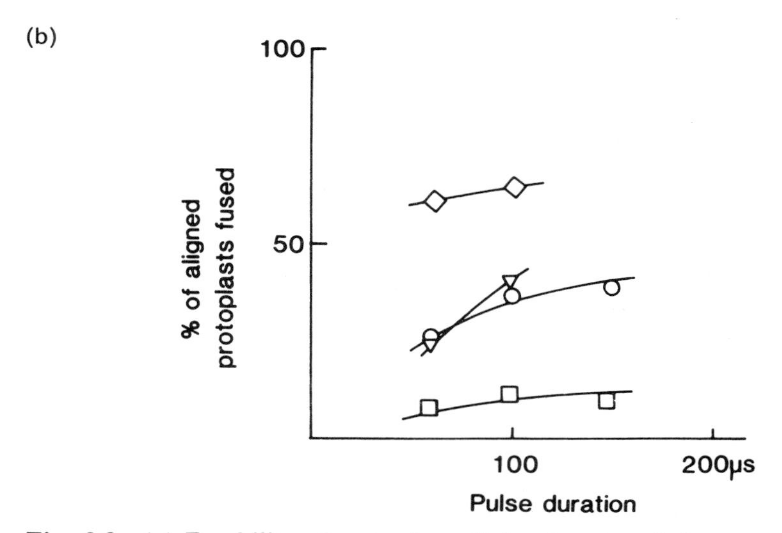

Fig. 6.6 (a) Fusability of protoplasts. Pulse duration–fusion response curves for self-fusion of leaf protoplasts of *Solanum brevidens*, cell suspension culture protoplasts of *S. brevidens* and *Datura innoxia*. (From Tempelaar and Jones, 1985, with permission.) (b) The effect on fusion frequency of pretreatment of protoplasts from *S. tuberosum* (dihaploid) cell suspension cultures. Control (0.5 M mannitol) (□); spermine pretreatment (∇); fusion in presence of 1 mM Ca Cl_2 (○); spermine pretreatment and presence of 1 mM $CaCl_2$ (◇). (From Tempelaar *et al.*, 1987, with permission.)

(1) The length of the chains of protoplasts—longer chains give higher fusion frequencies that short chains.
(2) Pulse length and voltage—longer pulses at higher voltages lead to more multifusion products.
(3) The fusion medium—inclusion of ions in the fusion medium can increase the percentage fusion.

The limitation here is that dielectrophoretic alignment only works in relatively non-conductive media, and so the usual fusion medium for plant protoplasts is mannitol plus 1 mM $CaCl_2$. Figure 6.6b shows that it is possible to increase the fusion frequency of less fusable protoplasts (e.g. small *S. tuberosum* cell culture protoplasts) with 1 mM $CaCl_2$, and by pretreatment with spermine. This pretreatment increases the area of membrane contact in the aligned chains. When both treatments are applied, the fusion frequency can be increased to 60%. Using this kind of approach it is possible to fuse all plant protoplasts examined in this way at relatively high frequencies. For practical purposes the pulse duration and voltage that give the highest fusion value for the shortest pulse duration are used. Protoplasts can, of course, be killed by too long pulses at too high voltages.

Other factors that may modulate the fusion frequency include: the osmotic pressure of the fusion medium—lower osmotic pressures lead to higher fusion frequencies, but at the expense of longer term viability; the enzymes used to isolate protoplasts may affect the response—some have protease contaminants which may alter membrane properties; and metabolic activity and protoplast type can also affect membrane composition and properties.

Heterofusions and Directed Fusion

When two populations of different protoplasts are mixed, the most fusogenic protoplasts will fuse to other types in the same way that they fuse to each other. It appears as if the most fusable protoplasts will initiate fusion with less fusable types. This provides a method of directing fusion towards heterokaryon formation. More fusable protoplasts are mixed with less fusable protoplasts at a ratio of about 1 : 5 to 1 : 10. When aligned, most of the more fusable protoplasts are in contact only with those of the less fusable type and not with each other; less fusable protoplasts will be in contact with each other. By selecting pulse voltage and durations that promote only the fusion of the more fusable protoplasts to their neighbours, most of the fusion products will be heterokaryons.

An alternative way of directing heterokaryon formation is to apply harsh fusion conditions, which kill the more fusable protoplasts but which allow the heterokaryons and more robust (less fusable) protoplasts to survive. Heterokaryons can then be purified by density-gradient centrifugation.

The production of heterokaryons at a relatively high frequency has been achieved for a range of plant protoplast types, and has so far not been found to be limited by protoplast size or origin. However, if the protoplast populations are very different in size, it may be necessary to reduce the voltage of the a.c. collecting field to prevent the smaller protoplasts being squeezed laterally out of the chains.

(a)

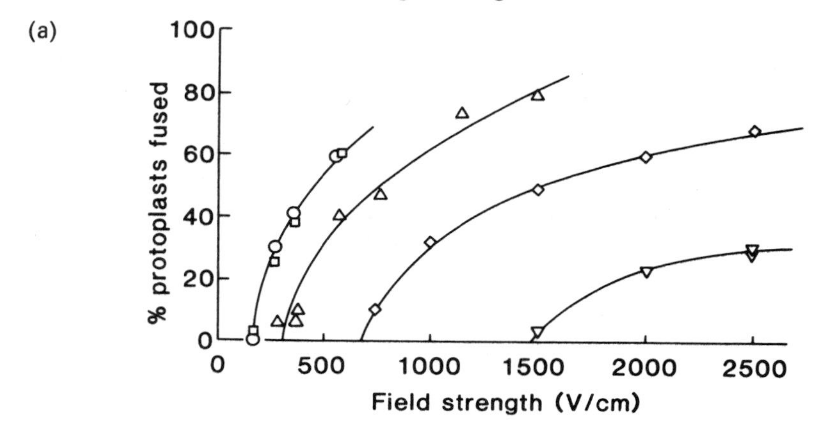

(b)

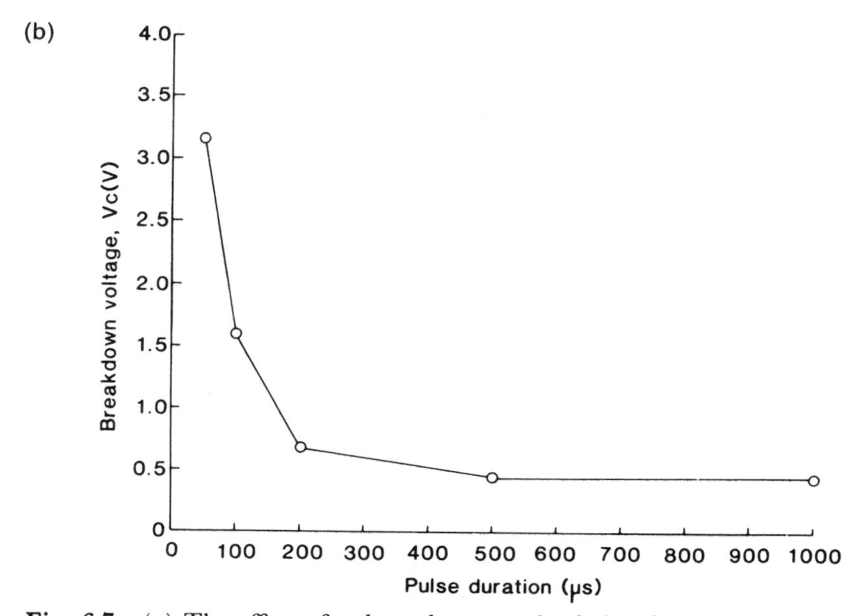

Fig. 6.7 (a) The effect of pulse voltage on the fusion frequence of dihaploid *S. tuberosum* protoplasts for pulses of different duration. Pulse durations: 50 μs (∇); 100 μs ($\Diamond$); 200 μs (Δ); 500 μs ($\bigcirc$); 1000 μs ($\Box$). The shorter the pulse duration, the higher is the field strength required before fusion occurs. (b) Relationship between the threshold membrane breakdown voltage (V_c) (of the same protoplasts as (a)) and pulse duration. (From Jones *et al.*, 1987, with permission.)

Electrofusion and Membrane Properties
The fusion pulse applied to protoplasts as described above causes reversible membrane breakdown, commonly referred to as 'pore' formation. Electrofusion can be applied to examine membrane properties. For example, the minimum (threshold) voltage at which pore formation (as monitored by fusion) occurs differs depending on the duration of the fusion pulse that is applied (Fig. 6.7a). Using these data, the transmembrane voltage (across the plasma membrane of a single protoplast) can be calculated, and is shown in Fig. 6.7b. It is clear that short pulses (in the microsecond range) must be applied at a higher voltage than longer pulses ($>200\,\mu s$) to achieve pore formation. This information is also relevant to the associated technique of 'electroporation' (see later) by which DNA (and other molecules) can be introduced into protoplasts directly.

MACROFUSION AND MICROFUSION

'Macrofusion'
The optimum conditions found using analytical electrodes can be directly used in scaled up fusions, with only a slight reduction in fusion frequency (caused by local circulation of the medium resulting in slight misalignment of the chains). Using five parallel stainless-steel electrodes inserted into a perspex well of volume 0.5 ml, 0.5 to 1×10^6 protoplasts can be fused in a cycle lasting 60–90 sec. Alternative electrode designs and flow cells can also be used. In a 1 : 1 mixture of two protoplast populations, at an overall fusion frequency of about 60%, 50% of the fusion products can be 1 : 1 ('binary') fusions, and, for protoplasts of similar fusability, half of these will be heterokaryons.

The next step is to culture the fusion products and to identify the hybrids. Bulk electrofusion is well suited for analysis with fluorescence-activated cell sorters (FACS), although plant protoplasts are technically more difficult to sort with FACS machines than animal cells. However, with the high fusion frequencies obtainable by electrofusion, complicated selection techniques are not necessarily needed, and it is possible to grow all the cells after fusion and to identify hybrid plants after shoot regeneration.

'Microfusion'
Macrofusion can be achieved simply and reproducibly, but yields a mixture of parental and hybrid cell types after fusion. An alternative but technically more demanding approach has been elegantly applied to plant protoplasts with the development of electrofusion of pairs of protoplasts. Its advantage is that hybrid cells produced after fusion need no further selection, and may be regenerated to somatic hybrid plants. However, this does require development of microculture techniques because plant protoplasts usually do not grow in isolation. This can be achieved by microdroplet culture—one protoplast in 50 nl medium is equivalent to 2×10^4 cells/ml in mass culture. The procedure developed to achieve microfusion is shown in Fig. 6.8.

This approach can be used for whole protoplast and subcellular fusions with enucleated protoplasts to transfer cytoplasmic traits, or with 'cytoplasts' (essen-

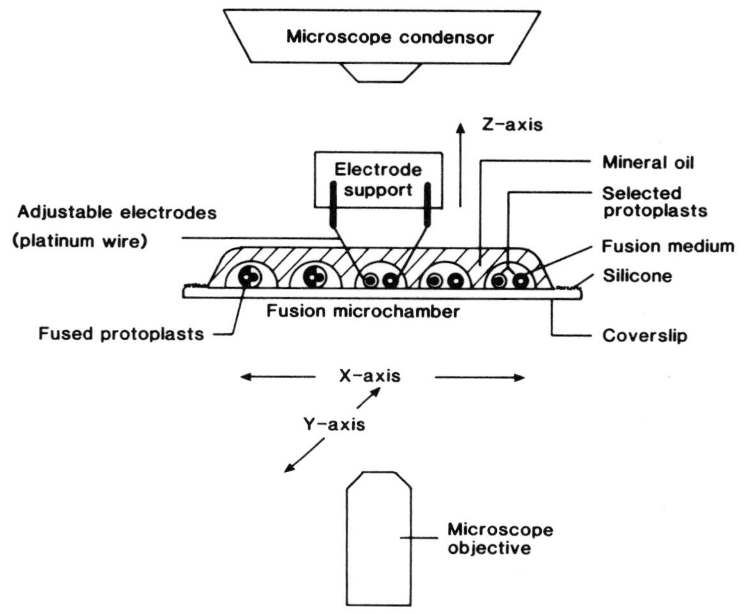

Fig. 6.8 'Microfusion' of individual protoplast pairs. Platinum electrodes (50-μm diameter), with a suitable gap between them, are fixed to an electrode support mounted under the condenser of an inverted microscope. Pre-selected protoplasts are transferred to a droplet of fusion medium under mineral oil. The microelectrodes are positioned on either side of the protoplasts, the electrofusion fields applied, and resulting heterokaryons can be transferred to droplets of culture medium for further culture. (Reproduced with permission from G. Spangenberg, Ph.D. thesis, University of Heidelberg, 1986.)

tially a nucleus surrounded by plasma membrane). Complete genome fusions yield in the order of 45% viable (dividing) heterokaryons, and sub-protoplast fusions in the order of 30% viable products. The fusion rate achieved is about 30 pairs of protoplasts fused and transferred to culture per hour. It is worth noting that this microsystem can also be used for microinjection of protoplasts, or, for example, a combination of nuclear injection of one protoplast followed by electrofusion to a second protoplast.

The control, efficiency and versatility of electrofusion, as indicated above, makes electrofusion the method of choice for fusion of plant protoplasts.

Identification of Somatic Hybrids

If heterokaryons have been isolated, either by micropipette or by using a cell sorter, then they may be cultured to yield hybrid plants. However, in most cases, a mixture of fusion products and parental types is obtained following fusion. Various approaches have been used to identify hybrid material, including the following:

(1) *Complementation selection of mutants*. Various types of mutants are available and can be used for selection. These may be antibiotic- or herbicide-resistant lines, auxotrophic mutants or non-allelic mutants of the same enzyme (e.g. *nia* and *cnx* mutants of nitrate reductase). The general principle is that if mutant A will grow on medium A′ (containing selectable agent A′ or required nutrient A′), and mutant B will grow on medium B′ (containing selectable agent B′ or required nutrient B′), then only somatic hybrids A + B should grow on a medium with both selectable agents A′ + B′.
(2) *Complementation of albino nuclear mutants*. Somatic hybrid callus of two non-allelic (nuclear) albino lines should complement to give green colonies/plants, which can be picked out of the mixture of albino and green colonies after fusion and culture.
(3) *Cultural differences*. Known differences in culture media requirements of parental protoplast types can be used either to enrich or to select hybrid colonies.
(4) *Fluorescence-activated cell sorters (FACS)*. Cell sorters have been used routinely to separate animal cell types by first labelling populations of cells with different fluorescent markers such as fluorescein or rhodamine. Cells travel individually in a liquid stream past lasers and photocells that detect the fluorescence, and different cell types are directed electrostatically into different containers. This approach has proved to be technically difficult to adapt to plant protoplasts, but is now being used to good effect. Usually the red autofluorescence of chlorophyll in leaf protoplasts is used as one marker, with a second population of colourless protoplasts labelled with fluorescein isothiocyanate as the other marker. Dual-labelled heterokaryons are detected and collected.
(5) *Mechanical isolation of individual heterokaryons*. A simple glass micropipette, attached to a syringe with a screw control, can be used to suck up individual heterokaryons whilst viewed under a light microscope. Heterokaryons between two protoplast types (e.g. leaf + cell suspension protoplasts) can easily be identified visually.
(6) *Mass culture*. Particularly with high fusion frequencies obtained by electrofusion, no selection may be applied early on after fusion. All the colonies obtained after fusion are cultured, and hybrid shoots may be detected by isozyme analysis or morphological markers (see below).

Characterization of Somatic Hybrids

Putative somatic hybrids may be confirmed and characterized in a number of ways, depending very much on the characteristics of the parental material. Approaches include:

(1) *Isoenzyme analysis*. If the parental material shows different bands on gels after staining for a particular isoenzyme, the hybrid usually expresses the sum of both patterns of bands, perhaps with additional bands derived from new combinations of enzymatic subunits. Isoenzymes commonly employed include glucose-6-phosphate dehydrogenase, phosphoglucose isomerase, glutamine oxaloacetate transaminase, esterase and shikimate dehydrogenase.

(2) *Nuclear DNA analysis*. If the DNA of parental material is extracted, fragmented with restriction enzymes, separated on a gel, transferred to a membrane and probed with a suitable [^{32}P]-labelled DNA fragment, autoradiograms will show different patterns of hybridization bands. Somatic hybrids show the sum of the two parental band patterns.

(3) *Enzyme activity*. If two complementing mutants, for example of nitrate reductase deficient (NR^-) lines are used, both parental lines cannot grow on nitrate as a N-source, as both have no nitrate reductase activity. Hybrids will grow on nitrate as the N-source, and the enzymatic activity of nitrate reductase can be measured.

(4) *Morphology*. Somatic hybrids may exhibit distinct morphological structures or pigments. These may be the sum of those present in both parents (e.g. leaf hairs from one parent and anthocyanin pigments from the other), or completely new characters. An example of the latter in potato is the presence of dark purple pigmentation in hybrid tubers obtained from parental lines, respectively with yellow and red tuber flesh.

(5) *Chloroplast DNA analysis*. Chloroplast DNA is extracted, fragmented and separated by gel electrophoresis. Different patterns of DNA fragments, diagnostic of the parental sources, can be obtained. Usually only chloroplasts from either one or other parental protoplast type are found in individual hybrids, so some sorting out of chloroplasts occurs during culture, with elimination of one chloroplast population.

(6) *Mitochondrial DNA analysis*. Mitochondrial DNA, extracted and analysed as above, usually reveals that recombination of microchondrial genomes has occurred in hybrids, so that banding patterns of DNA unique to individual hybrids are obtained.

(7) *Cytological analysis*. The chromosome complement of somatic hybrids should be determined routinely by root-tip squashes. Such analyses will reveal if the hybrids possess the expected complement of chromosomes from each parent, whether more than two protoplasts have fused together, the degree of aneuploidy, and also perhaps the extent of exchange between different chromosomes. The latter aspect can be analysed further using techniques of *in situ* hybridization.

(8) *Analysis of desired characters*. The combination of useful agronomic characters (e.g. pathogen resistance) can be assessed by standard breeders' tests.

Fig. 6.9 *Left: S. tuberosum* parent (24 chromosomes); *right: Solanum brevidens* parent (24 chromosomes); *middle:* tetraploid somatic hybrid (48 chromosomes) that expresses viral resistance from the *S. brevidens* parent.

Some Applications of Protoplast Fusion

Although a series of somatic hybrid plants of model plant species such as tobacco has been produced, the major practical applications are in genetic manipulation of crop plants. The most obvious approach, which has been successful in, for example, potato, is the introduction of useful agronomic characters from sexually incompatible wild species into crop plant germplasm.

FUSION OF POTATO AND *SOLANUM BREVIDENS*: A CASE STUDY OF SOMATIC HYBRIDIZATION

The wild potato species *S. brevidens* is resistant to potato leaf roll virus (PLRV) and potato virus Y (PVY), two important viral diseases of commercial potato (*S. tuberosum*). However, *S. brevidens* cannot be sexually crossed directly with *S. tuberosum*. Somatic hybrids, produced by protoplast fusion using both chemical and electrofusion (Fig. 6.9), can exhibit resistance to PLRV and PVY from the wild parent. The range of genetic combinations found in these hybrids is summarized in Table 6.3. In this case, the chemical fusion method (Ca^{2+} at high pH) was less efficient than electrofusion; with the latter technique 12.6% of regenerated shoots (without prior selection) were hybrid. The parental potato

Table 6.3 Genetic combinations and properties of somatic hybrids between *S. tuberosum* and *S. brevidens*

Fusion method	*Chemical*	*Electrical*
Number of hybrids	11	49
Percentage of regenerated shoots that were hybrid	2.3%	12.6%

Cytology
Euploid
tetraploid (48) 20%
hexaploid (72) 14%
Aneuploid
tetraploid 23%
hexaplopid 34%
octaploid 9%

Chloroplast origin
Solanum tuberosum 62%
Solanum brevidens 38%

Classes of euploid hybrids:

	Nuclear genome dosage		*Chloroplast origin*	
	S. tuberosum	*S. brevidens*	*S. tuberosum*	*S. brevidens*
Tetraploids (48)	1	1	1	–
	1	1	–	1
Hexaploids (72)	1	2	1	–
	2	1	1	–
	1	2	–	1
	2	1	–	1

genotype was dihaploid (i.e. 24 chromosomes), and the *S. brevidens* also possessed 24 chromosomes. Euploid hybrids with 48 and 72 chromosomes were obtained (products of two or three protoplasts fusing together), and also a range of aneuploids. More hybrids retained the *S. brevidens* chloroplast type than that of potato, and all the genetic combinations indicated in Table 6.3 were found. Tests for resistance to PLRV and PVY showed that some hybrids exhibited a similar level of resistance as the *S. brevidens* parent. An important point is that many of the hybrids were female fertile at least and could be crossed with potato varieties, so allowing introgression of the useful virus resistance genes into potato germplasm. Hybrid plants have been evaluated in field trials.

In Japan, somatic hybrids of *S. tuberosum* and *Lycopersicon pimpinellifolium*, a wild tomato species with resistance to late blight, bacterial wilt and high temperatures, have been evaluated in the field. It is also possible to combine selected potato lines directly (at the reduced dihaploid ploidy level) by fusion to synthesize new tetraploids that may be of direct agronomic value.

PARTIAL GENOME TRANSFER

Other areas of application include partial genome transfer from an irradiated 'donor' protoplast by fusion to an acceptor protoplast and transfer of cytoplasmically encoded traits (i.e. in mitochondria and chloroplasts). An example of the former is the production of asymmetric hybrid plants after electrofusion of irradiated protoplast of kanamycin-resistant *Nicotiana plumbaginifolia* with *N. tabacum*. In this case, following irradiation of the donor protoplasts and fusion with *N. tabacum*, plants that were morphologically similar to *N. tabacum* were regenerated and could be classified into three types: (1) somatic hybrids with a reduced contribution from *N. plumbaginifolia* (kanamycin resistant, with one or two copies of *N. plumbaginifolia* esterase isoenzymes); (2) asymmetric hybrids (but with no copies of *N. plumbaginifolia* esterases); and (3) 'cybrids' (kanamycin sensitive) in which the *N. plumbaginifolia* cytoplasm alone has been transferred to a *N. tabacum* nuclear background. The last category of plants were male sterile, and this type of cytoplasmic male sterility is of considerable practical interest to plant breeders for the production of F_1 hybrid seeds of various crop plants.

Cell-fusion techniques thus allow the transfer or combination of identifiable agronomically-useful traits without the need for detailed molecular information, and also enable transfer of polygenic factors that may underlie such traits. With improved fusion procedures, protoplast regeneration protocols and further understanding of the control of genetic stability of plant cells in culture, many more examples of the practical applications of this approach can be expected.

Specific Gene Transfer into Plant Cells: 'Transformation'

In contrast to the combination of complete or partial genomes by protoplast fusion, it is also possible to introduce one or a few genes directly into plants by technology known as 'specific gene transfer'. The aim is to introduce new characters encoded by such genes directly into existing crop varieties, for example, thus retaining all the beneficial properties of the variety and avoiding the genetic reassortments that occur during normal plant breeding procedures. As has already been discussed in Chapter 3, it is also possible to introduce genes whose products are only expressed in specific organs of the plant, or at specific times in development. The methods used to transfer specific genes in this way, with the production of 'transgenic' plants, can be divided into two broad categories—indirect gene transfer using 'vectors', and direct gene transfer. These approaches are summarized in Table 6.4 and are considered below.

Agrobacterium*-mediated Gene Transfer*

Agrobacterium tumefaciens, and the closely-related species *A. rhizogenes*, are soil-borne bacterial pathogens of plants. *Agrobacterium tumefaciens*, also known as 'crown gall bacterium', enters damaged plant tissues and causes cell proliferation at the wound site, resulting in the formation of a gall or tumour. *Agrobacterium rhizogenes* similarly infects wounded tissues, but in this case induces the production of roots, and is the causative agent of 'hairy root disease'.

Table 6.4 Methods of specific gene transfer

Dicotyledons	*Monocotyledons**
A. Indirect gene transfer	
• *Agrobacterium tumefaciens*	• 'Agroinfection'
• *Agrobacterium rhizogenes*	• ? pollen/DNA transfer
• DNA viruses	
B. Direct gene transfer	
• gene transfer into protoplasts (chemically or electrically mediated)	
• microinjection of DNA into nuclei	
• fusion of protoplasts with DNA-containing liposomes	
• 'shotgun' transformation	
• 'macroinjection'	
• ? seed soaking	

*Particularly Gramineae.

The mechanism underlying this process is illustrated in Fig. 6.10. On contact with wounded plant cells, a segment of DNA from the Ti (tumour-inducing) plasmid of *A. tumefaciens* is transferred from the bacterium to the plant cell, where it becomes integrated into one or more of the plant cell's chromosomes. This transferred or 'T'-DNA contains so-called '*onc*' genes, whose expression results in higher endogenous levels of plant growth regulators, auxins (indoleacetic acid) and cytokinin (zeatin riboside), and these in turn stimulate cell divisions that lead to tumour formation. This process occurs naturally as part of the normal infective process of these bacteria in soil, but has been exploited by molecular biologists as a vector system to introduce specific genes into plant cells.

The T-DNA also encodes several genes responsible for the synthesis of compounds called 'opines' (i.e. nopaline, octopine, mannopine), which are metabolic substrates for the bacteria and can be the sole carbon and nitrogen source required for their growth. This elegant bacterial genetic engineering diverts the plant cell metabolism to producing compounds (i.e. opines) which cannot be metabolized by the plant, but on which the bacteria can proliferate.

The Ti plasmid of wild type *Agrobacterium* is relatively large (about 200 kilobases) and difficult to manipulate experimentally. It was found that a region of the Ti plasmid outside the T-DNA, referred to as the virulence region, carries genes (*vir* genes) that are involved in the transfer of the T-DNA into plant cells (Fig. 6.11a). Although the precise mechanism of T-DNA transfer is not understood, molecular analysis of T-DNA integrated in host chromosomes has shown the presence of 25 nucleotide directly-repeated sequences flanking the boundaries of the integration sites. Genes that are to be introduced into plant cells must be inserted between these 'left' and 'right' border sequences (Fig. 6.11), or next to at least one border of T-DNA. Thus by cloning foreign genes into the T-DNA of the Ti plasmid, it is possible to exploit the natural ability of *Agrobacterium* to transfer new DNA into the plant genome.

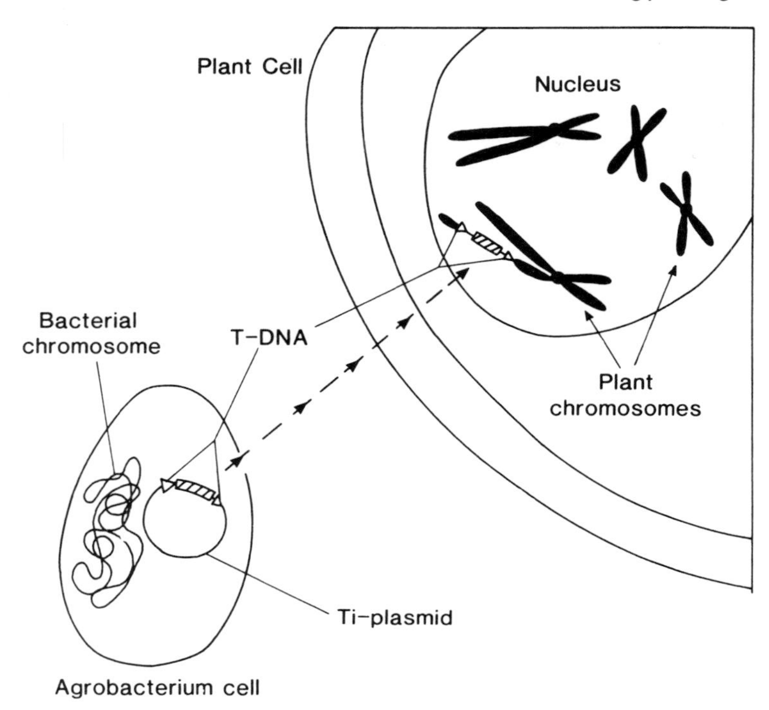

Fig. 6.10 T-DNA transfer from *Agrobacterium* to a plant cell (courtesy of Dr D. Twell).

Plant tissues transformed in this way by wild-type *Agrobacterium* can be identified by their characteristic tumorous growth, caused by the excess of growth regulators and their ability to proliferate on hormone-free medium. However, such tissues, which also produce opines, are incapable of regenerating normal plants, and are therefore of no practical use. It has therefore been necessary to remove the oncogenic properties of the T-DNA, but retain the border sequences to produce 'disarmed' Ti plasmids. As there is no excess production of plant hormones, whole plants transformed with such disarmed *Agrobacterium* strains can be produced and detected by the produciton of opines. A more effective method is to introduce a selectable marker gene, such as one encoding for resistance to antibiotics such as kanamycin. Transformed cells can then be selected by their ability to grow on media containing the selectable antibiotic, whilst untransformed cells do not (Fig. 6.12). However, the large size of the Ti plasmid still makes this transformation vector difficult to manipulate.

A significant advance that bypasses the problem of Ti plasmid size was the discovery that the T-DNA and the *vir* region could be separated on to two different plasmids without loss of the T-DNA transfer capability, i.e. they worked in a *trans*

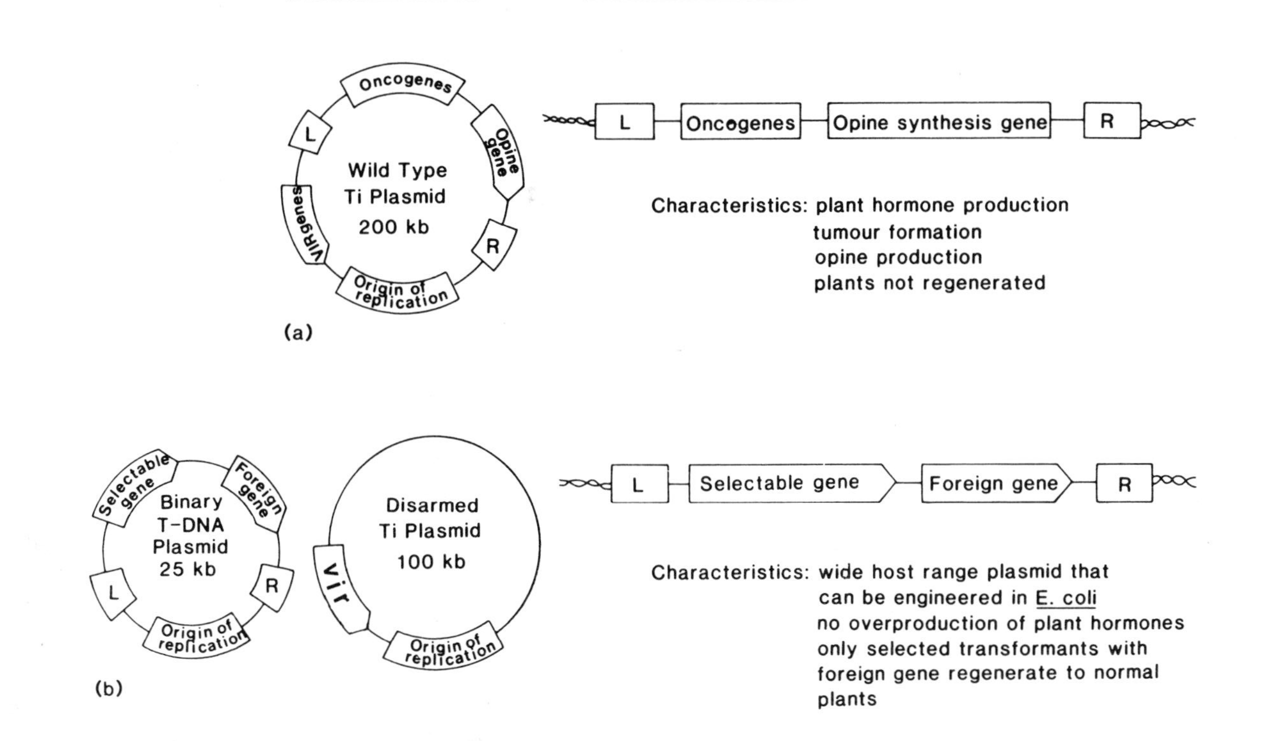

Fig. 6.11 Ti plasmid-mediated T-DNA transfer. (a) Ti plasmid of wild *Agrobacterium*. L, left T-DNA border sequence; R, right T-DNA border sequence. (b) Binary T-DNA plasmid and 'disarmed' Ti plasmid, in which the *vir* genes of the Ti plasmid act in trans to transfer T-DNA, with selectable and foreign gene, to host chromosomes.

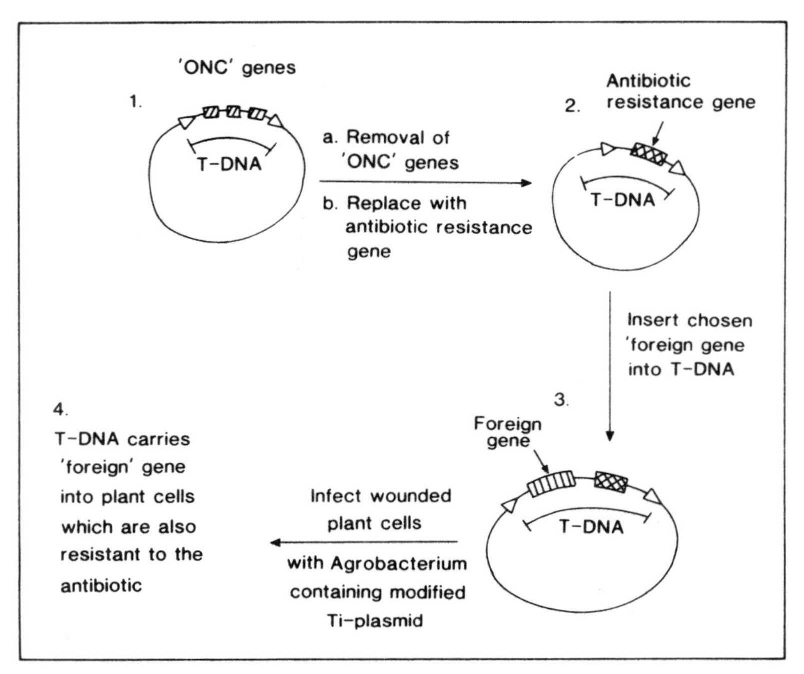

Fig. 6.12 Modification of Ti plasmid to create a 'disarmed' vector with selectable marker gene giving resistance to an antibiotic (courtesy of Dr D. Twell).

as well as a *cis* configuration. This discovery led to the development of 'binary' T-DNA vectors that involve two plasmids (Fig. 6.11b)—a disarmed Ti plasmid containing the *vir* genes, and a separate smaller plasmid containing the T-DNA (Bevan, 1984). The small binary T-DNA plasmid is a wide host range plasmid that can replicate both in *E. coli* and in *Agrobacterium* cells. The desired foreign gene is inserted into the binary T-DNA plasmid between the left and right border sequences. A selectable plant marker gene is also inserted and, in combination with the transfer of the desired foreign gene, allows selection of transformed plant material. A list of agricultural plants transformed in this way via *Agrobacterium* is given in Table 6.5 and commonly used selectable and marker genes in Table 6.6.

Methods of Agrobacterium *Transformation*

Agrobacterium-mediated gene transfer methods are being developed for a wide range of dicotyledonous plants and gymnosperm species. Normally bacteria are incubated with plant cells for a few hours to 1–2 days, during which time T-DNA transfer takes place. The cells are then washed and treated with antibiotics to remove the bacteria, which would otherwise over-run the cultures. The cells are then cultured in the presence of the selectable agent, and transformed shoots are regenerated and characterized.

Table 6.5 Agricultural plants transformed via *Agrobacterium*

Apium graveolens (celery)	*Latuca sativa* (lettuce)
Asparagus officinalis (asparagus)	*Linum perenne* (flax)
Beta vulgaris (sugar-beet)	*Lotus corniculatus* (birds-foot trefoil)
Brassica napus (oilseed rape)	*Lycopersicon esculentum* (tomato)
Brassica rapa (turnip)	*Medicago sativa* (alfalfa or lucerne)
Cucumis sativus (cucumber)	*Nicotiana* spp. (tobacco species)
Daucus carota (carrot)	*Petunia* spp. (petunia species)
Glycine max (soybean)	*Phaseolus vulgaris* (French bean)
Gossypium hirsutum (cotton)	*Populus trichocarpa* × *deltoides* (poplar)
Helianthus annuus (sunflower)	*Solanum tuberosum* (potato)
Juglans regia (walnut)	*Trifolium* spp. (clover species)
	Vigna unguiculata (cowpea)

Table 6.6 Dominant selectable marker and reporter genes

Marker gene	*Enzyme encoded*	*Resistance conferred*
Antibiotics		
npt II	neomycin phosphotransferase	kanamycin neomycin G418 paromomycin
hpt or *aph IV*	hygromycin phosphotransferase	hygromycin
dhfr bacterial or mouse	dihydrofolate reductase	methotrexate
Herbicides		
bar	phosphinothricin acetyltransferase	phosphinothricin
aro A	5-enolpyruvylshikimate-3-phosphate synthase	glyphosate
modified *als* genes	acetohydroxyacid synthase (or acetolactate synthase)	chlorsulfuron imidazolanones
Reporter gene		
CAT	chloramphenicol acetyl transferase	
GUS	β-glucuronidase	
npt II	neomycin phosphotransferase	
luc	luciferase	
bar	phosphinothricin acetyltransferase	
β-gal	β-galactosidase	

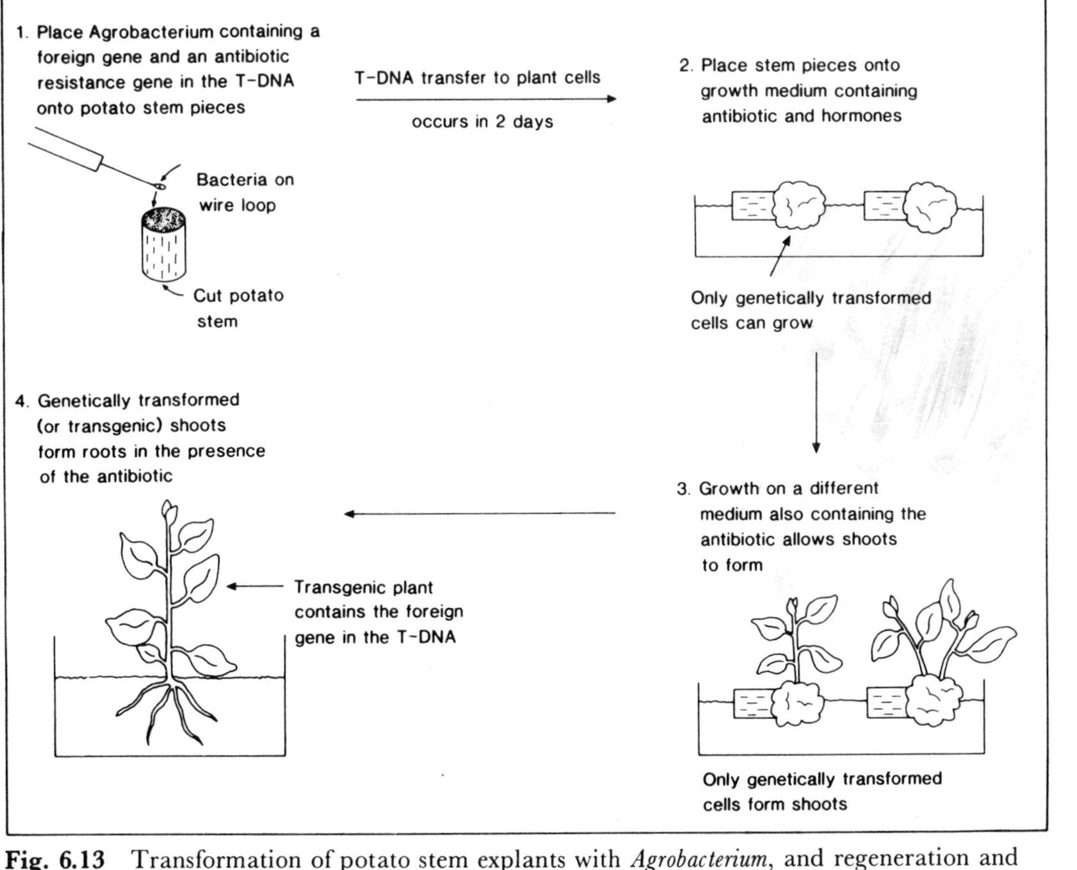

Fig. 6.13 Transformation of potato stem explants with *Agrobacterium*, and regeneration and selection of genetically transformed shoots and plants (courtesy of Dr D. Twell).

There are two main transformation strategies: inoculation of explants (stem pieces, leaf discs etc.) or of protoplasts.

Explant transformation The most widely used approach for *Agrobacterium*-induced transformation is to culture bacteria with stem segments (Fig. 6.13), leaf discs or other suitable explants. The choice of tissue is usually governed by the ease of plant regeneration from somatic cells of the explant. In the case of leaf disc explants, which can be dipped into a solution of *Agrobacterium*, a 'feeder layer' of dividing suspension culture cells is often incorporated in the agar medium and serves to stimulate division and growth of *Agrobacterium*-treated cells. When *A. rhizogenes* is used as a gene vector, then inoculation of stem punctures (with a sterile needle) is usually the method chosen. This is because transformed roots, containing *A. rhizogenes* T-DNA, readily grow out from such inoculation sites. Transformation is stimulated by and normally occurs at wounded plant cells, both because bacterial access to plant cells is increased and because specific phenolic compounds (e.g. acetosyringone, syringone) stimulate the transfer process. The process of inoculation, selection and regeneration of transformed plants takes 1–3 months depending on the species and has been successfully achieved for most important dicotyledonous crop species.

Co-cultivation of Agrobacterium *with protoplasts* An alternative transformation procedure involves the co-cultivation of *Agrobacterium* with isolated protoplasts. The advantage of this over explant methods is that large numbers of protoplasts can be treated in each co-cultivation to give large numbers of independent transformation events. In addition, the cells can be exposed uniformly to the selective agent used and this should result in fewer 'escapes' (non-transformed material) after selection. However, the disadvantage is that protoplast co-cultivation is limited to those species and genotypes which can readily be regenerated back to plants, and protoplast procedures are technically more difficult than explant regeneration. Regeneration back to intact plants also takes longer from protoplasts.

Criteria for Selection

Since only a small proportion of cells are transformed by *A. tumefaciens*, it is important to have a selectable marker gene that enables transformed tissues to grow when non-transformed cells do not. The features of a good selectable marker are:

1. Transformed cells should exhibit a high level of resistance to the drug or herbicide that completely inhibits growth of non-transformed cells.
2. Easy detection and efficient selection of rare transformants from a large excess of surrounding inhibited wild-type cells should be achieved.

A list of commonly used dominant selectable marker genes is given in Table 6.7. Although the above criteria can be met for some combinations of species and marker genes, in many cases selection is not so clear cut. For example, some non-transformed shoots may grow in the presence of the selection agent, possibly as a

Table 6.7 Commonly used promoters

Constitutive promoters	*Source*
nos	nopaline synthase
ocs	octopine synthase
man	mannopine synthase
CaMV 19S, 35S	cauliflower mosaic virus
Tissue specific promoters	*Tissue expression*
α-amylase	aleurone
glutenin	endosperm
β-hordein	endosperm
patatin	tuber
phaseollin	cotyledon
conglycinin	cotyledon
Inducible promoters	*Induction conditions*
small subunit Rubisco	light
chlorophyll *a/b* binding protein	light
alcohol dehydrogenase-1	anaerobiosis
heat shock proteins	heat shock
chalcone synthase	fungal elicitors
potato inhibitor-II	wounding

result of cross-protection by transformed tissues. A second round of selection, for example testing for the ability to root in the selection agent, may be required. In some species (e.g. tomato) the explant is very sensitive to the bacteria, and a careful control of bacterial titre is required. In other cases (e.g. cereals) there is inherent resistance to some selectable agents, such as the aminoglycoside antibiotics, so the 'window' for effective selection is small. Different-sized explants are also differentially sensitive to selectable agents. Sometimes the marker gene may be stably integrated and expressed, but does not confer effective resistance to the selection agent at the concentration used. There may also be differences in the level of marker gene product expressed depending on the site of insertion of the gene in the plant chromosome (position effects), resulting in differential resistance. In any case the presence of marker gene DNA usually needs to be confirmed by Southern analysis.

Alternative Approaches—Direct Gene Transfer

Although it is relatively straightforward to introduce foreign DNA into cells of many dicotyledonous species using *A. tumefaciens* or *A. rhizogenes* as vectors, as described above, use of *Agrobacterium* as a vector has met with very little success when applied to the Gramineae and cereals in particular. Since cereals are by far the most important agricultural crops worldwide, this fact has led to a variety of different approaches for transformation being examined.

An important advance in this area of 'direct gene transfer' is based on the demonstration that, as for animal or bacterial cells, a vector to carry in and aid integration of foreign DNA is not, in fact, necessary. For plants, this was demonstrated by the stable transformation of protoplasts by direct introduction of DNA constructs that contained selectable marker genes. Using either chemical methods (usually based on polyethyleneglycol treatment) or electrical methods ('electroporation', see below), or a combination of physical and chemical treatments, relatively high protoplast transformation frequencies were obtained from tobacco protoplasts (1–10% of treated protoplasts), with subsequent regeneration of whole transformed plants. Fusion of protoplasts with DNA-containing liposomes can also result in stable transformation. Soon afterwards, it was demonstrated that direct gene transfer could be applied equally to protoplasts of gramineous species such as *Triticum monococcum*, *Zea mays* (maize) and *Lolium multiflorum* (Italian ryegrass). In other words, there was no molecular block to cereal transformation—rather it is a cell biological problem of devising methods to introduce foreign genes into cereal cells capable of regenerating intact plants. Unfortunately as we have already noted, with the exception of rice and perhaps also maize, the major cereals cannot as yet be regenerated routinely to plants from protoplasts.

Alternative approaches to the use of *Agrobacterium* as a DNA vector are now considered in more detail, with particular reference to cereal transformation.

Cereal Transformation and Direct Gene Transfer

Although the first two sections below, which are based on *Agrobacterium* studies, are not methods of *direct* gene transfer, they are discussed here for the sake of completeness in relation to approaches to transformation of cereals.

1. *Agrobacterium* Cereals do not respond to *Agrobacterium* infection in the same way as dicotyledons. The reason for this is not clear, but it may be because bacteria do not attach to cereal cell walls, or because hormonal responses are different in cereals, or because specific wound substances are absent or present at too low concentrations at wound sites to promote regions on the Ti plasmid (*vir* genes) to stimulate the transfer process. Stable integration of T-DNA has recently been demonstrated for the monocotyledon *Dioscorea bulbifera* (yam) following pre-incubation of the bacteria with wound substances from potato. A similar approach could work for cereals. There is also evidence that T-DNA may be introduced into maize cells by *Agrobacterium*, although evidence for stable integration of genes is lacking. The latter observation is supported by another approach, considered next.

2. '*Agroinfection*' This approach combines *Agrobacterium* transfer and viral infection. DNA geminiviruses (e.g. maize streak virus, MSV; and wheat dwarf virus, WDV) can be introduced into the T-DNA region of *Agrobacterium* vectors, and following infection of a wound site, presence of the virus in the plant provides a very sensitive assay that DNA from the vector has been introduced into the host plant (e.g. Grimsley *et al.*, 1987). This has been achieved for MSV in maize and WDV in wheat. Control experiments have shown that naked MSV DNA is itself not infectious, and that *Agrobacterium* can introduce MSV DNA into cereals. However, so far there is no evidence for stable integration of DNA, and the MSV infection may originate from a low frequency of transfer events. Nevertheless, this

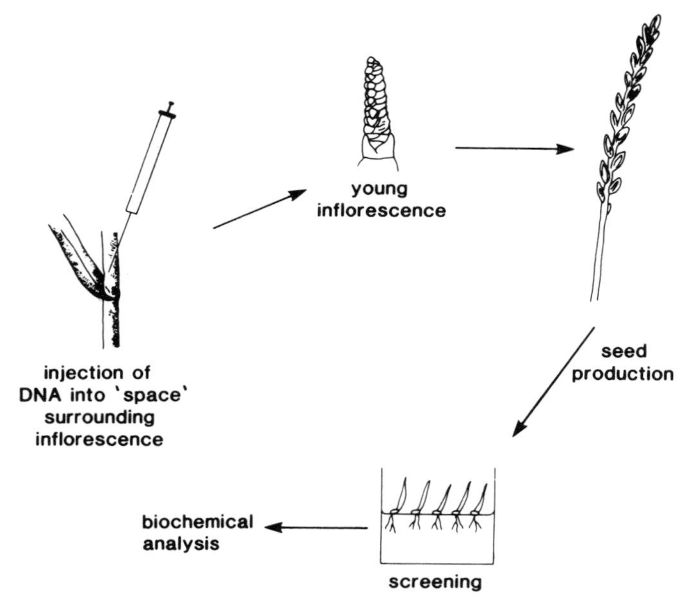

Fig. 6.14 'Macroinjection' of selectable DNA gene construct of developing rye inflorescence, with subsequent selection for transformed seeds (courtesy of Dr B.T. Lee).

approach will allow study of expression of genes of interest, engineered in the MSV (or WDV), to be studied in intact plants. The introduction of transposable elements into maize carried within MSV could also lead to stable integration within maize cells.

3. *Macroinjection* There has been a report that a simple technique has enabled the recovery of three rye plants containing a foreign gene conferring resistance to the antibiotic kanamycin. The procedure involved the injection of an aqueous solution of the kanamycin resistance gene (in a plasmid) into the space surrounding the developing inflorescences about 14 days before meiosis (de la Pena *et al.*, 1987; Fig. 6.14). The injected plants were allowed to produce seeds, which were screened for resistance to kanamycin. Stable integration of the resistance gene was demonstrated in resistant seedlings by Southern analysis. It remains to be seen whether this approach will have successful wider applications.

4. *Microinjection* Technically more demanding than macroinjection, microinjection of selectable DNA into nuclei of plant cells has not been fully exploited as an efficient system for transformation of dicotyledons (Fig. 6.15). Intranuclear injection of tobacco or alfalfa protoplasts (e.g. Reich *et al.*, 1986) can result in a transformation frequency of about 15–25%. The same approach is being attempted to inject nuclei of cereal egg cells or pro-embryoidal cultured cells (e.g. from cultured

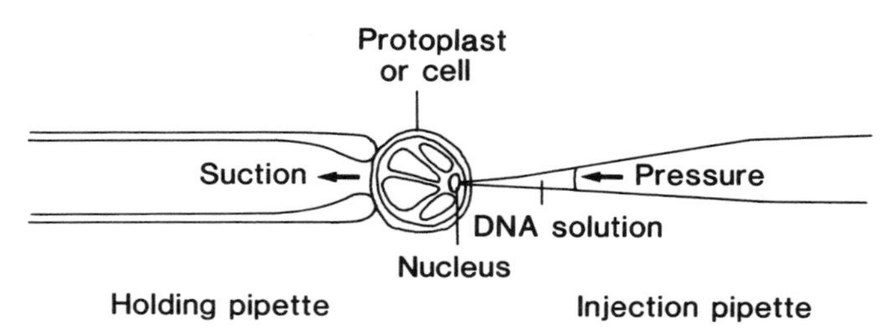

Fig. 6.15 Microinjection of protoplasts or cells. The cell is held by suction on a holding pipette. The injection pipette is inserted into the nucleus, and DNA solution injected.

microspores). Such cells are capable of developing to whole plants, but the problem has been to locate and inject the nucleus. Results from this approach applied to cereals are awaited with interest. In addition, it may be possible to transfer cell organelles or isolated chromosomes by microinjection.

5. *'Shotgun' transformation* A novel approach to try to introduce selectable DNA into cells in immature cultured explants (e.g. immature embryos) capable of regenerating to give somatic plants has been developed by Klein *et al.* (1987). This involves the use of a shotgun that fires tungsten microprojectiles (spherical particles, 4-μm diameter) using an explosive charge into cells. DNA or viral RNA can be pre-adsorbed on to the microprojectiles, and after firing them into epidermal strips of onion cells, about 90% of the cells contain the particles, and products of the introduced genes, or virus particles can be assayed. The approach is now being applied to morphogenetic cereal tissues, but results have yet to be reported.

6. *DNA/pollen transfer, seed soaking* Two approaches used, so far without molecular evidence of success, are 'DNA/pollen transfer' and 'seed soaking'. In the former, pollen and DNA containing a selectable gene are mixed and applied to the stigma. The rationale is that the DNA is carried by the pollen down to the egg cell, which it may enter during fertilization. In the second method, cereal-dehydrated embryos are isolated in organic solvent and imbibed in an aqueous DNA solution. During imbibition, some of the DNA may enter germ-line cells leading to transformation.

7. *Chemically-mediated DNA uptake* Following work on animal cells, a variety of methods has been tested to induce DNA uptake into protoplasts by chemical means. The most successful approach has been to use polyethylene glycol as such an agent. Successful introduction of foreign selectable DNA, followed by recovery of transformed cells, has been achieved for a number of cereals and grasses. In this and the following methods, excess 'carrier' DNA is usually added with the selectable gene to reduce degradation by nucleases and perhaps also to facilitate integration of the marker gene.

8. *Electroporation* As has been previously described in the section on electrofusion, when plant protoplasts (or animal cells) are exposed to a suitable external electric field the membrane insulation will break down, resulting in the development of pores ('electroporation'). If conditions are chosen correctly, then the induced membrane breakdown is reversible, and it can re-seal (Zimmermann, 1986). If foreign DNA (or other molecules) are present in the external solution during such treatment, they can enter the cells through the pores. Two basic types of electric field may be used – either rectangular pulses in which the pulse duration is precisely controlled, or a 'spike' discharge from a capacitor. For the purposes of introducing DNA, pulses are commonly either of short duration and high voltage, or longer duration and lower voltage. To maintain protoplast viability, rectangular pulses must be of relatively short duration (microseconds) because their value exceeds the membrane breakdown voltage throughout, whereas for a 'spike' pulse, in which the membrane breakdown voltage is only exceeded for a short period, the tail of the pulse can be much longer (milliseconds). It appears that DNA may enter cells via electrophoresis on the longer tail of capacitor discharge pulses following electroporation.

Electroporation has been used successfully to introduce selectable constructs and select cell lines that contain integrated foreign DNA (e.g. Fromm *et al.*, 1986). For example, a transformation frequency for maize of 1% has been obtained using capacitor discharge pulses. This approach therefore holds great promise as a method of obtaining transformed plants, but is at present limited to those gramineous species, such as rice, in which plants can be regenerated from protoplasts.

Transient Expression of Introduced Genes

As we have seen, foreign genes can be introduced into protoplasts by direct gene transfer mediated either chemically or by electroporation. Even without regeneration to transformed plants, some aspects of expression of gene constructs can be examined. To avoid the lengthy process of transformation, selection and bulking of tissue for analysis of integrated genes (which may take 3–5 months), it is possible to study 'transient expression' of the constructs introduced directly into protoplasts. The approaches are outlined in Fig. 6.16. Most conveniently, DNA constructs are added to protoplasts, introduced into them by electroporation or by chemical methods, after which the protoplasts are cultured for 1–2 days. The protoplasts are then harvested and assayed for products encoded by the introduced (but not necessarily integrated) genes. An example of expression of the CAT (chloramphenicol acetyltransferase) gene is given in Fig. 6.17. Information on aspects of gene constructs such as the relative level and tissue specificity of expression can be obtained. In the latter case it is necessary to isolate protoplasts from specific cell types, and this can now be accomplished for some tissues. There is uncertainty as to the length of time protoplasts isolated from specific tissues reflect the metabolic properties of those tissues, but this may be of the order of several days after isolation, depending on the conditions of culture and the protoplast source.

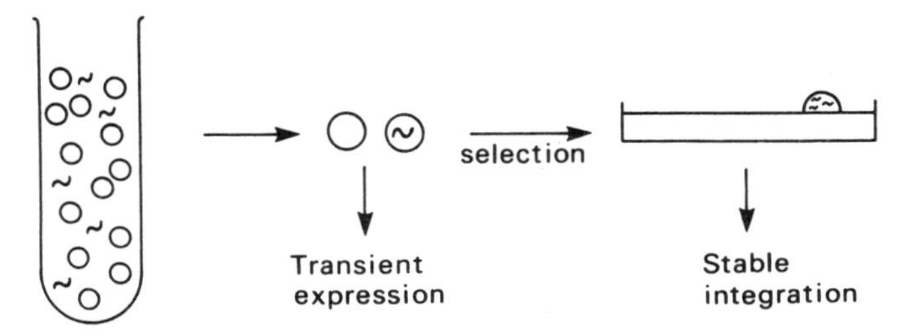

Fig. 6.16 DNA contrasts (represented by wavy lines), mixed with protoplasts, can be introduced into protoplasts by chemical treatments or electroporation. Transient expression of the encoded genes can be assessed after culture for 1–2 days, or stable integration examined after selection for growth of transformed cells in culture after 1–3 months.

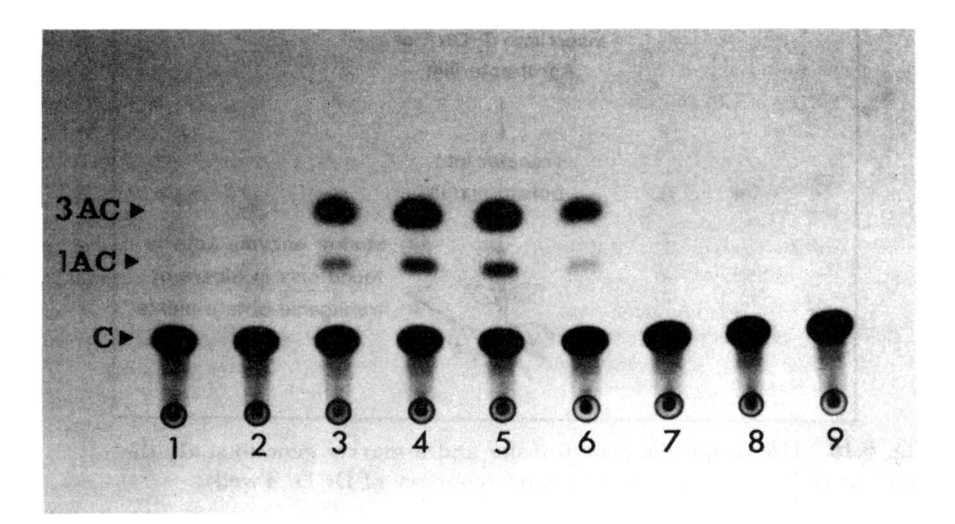

Fig. 6.17 Expression of the marker gene chloramphenicol acetyltransferase in electroporated leaf protoplasts of sugar beet, at a range of field strengths. Activity is optimum at 500–750 V/cm^{-1}, 3 pulses of 99.9 μs. C = chloramphenicol; 1AC = 1-acetylchloramphenicol; 3 AC = 3-acetylchloramphenicol. Tracks: 1 = negative control; 2 = 100 V cm^{-1}; 3 = 250 V cm^{-1};4 = 500 V cm^{-1};5 = 750 V cm^{-1};6 = 1000 V cm^{-1};7 = 1500 V cm^{-1};8 = 2000 V cm^{-1};9 = 2500 V cm^{-1}.

Use of Specific Gene Transfer to Study Gene Function

The potential applications of specific gene transfer are considerable, and are discussed in more detail in Chapters 7–10. We provide one example here, typical of many of the applications described, on the production and use of a chimaeric

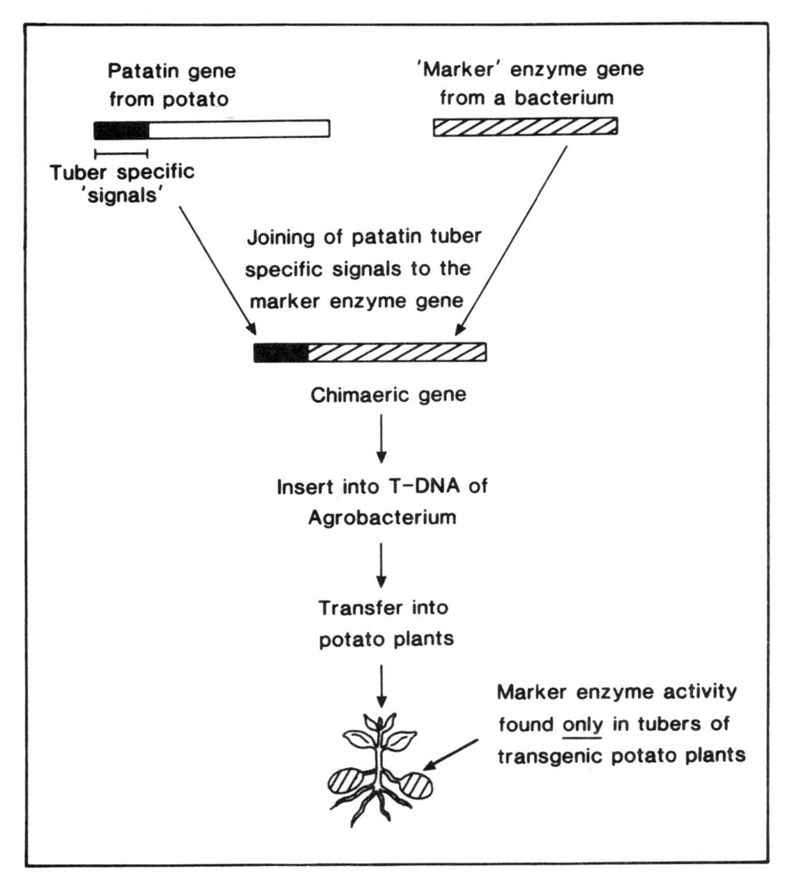

Fig. 6.18 Use of specific gene transfer and a marker gene to study the function of the patatin gene of potato (courtesy of Dr D. Twell).

gene construct to identify tuber specific signals of the most abundant tuber-specific protein called 'patatin'. As shown in Fig. 6.18, the patatin gene has a promoter region containing regulatory sequences which specify expression in potato tubers and the structural gene that codes for the patatin protein. If the structural gene is replaced by an easily assayed marker enzyme gene, to produce a 'chimaeric' gene, this can be inserted into *Agrobacterium* T-DNA and transferred to potato plants. The marker gene activity is then found only in the tubers of the transformed or 'transgenic' plants. Thus it is possible to direct expression of introduced genes to specific plant organs, and by modifying the promoter sequences, to alter their level of expression.

General Reading

Gleba, Y.Y. and Sytnik, K.M. (1984). *Protoplast Fusion.* Berlin, Springer-Verlag.
Jones, M.G.K. (1988). 'Fusing plant protoplasts', *Trends Biotechnol.* **6**, pp. 153–158.
Klee, H., Horsch, R. and Rogers, S. (1987). '*Agrobacterium*-mediated plant transformation and its further applications to biology', *Ann. Rev. Plant Physiol.* **38**, pp. 467–486.
Zimmermann, U. (1982). 'Electric field-mediated fusion and related electrical phenomena', *Biochim. Biophys. Acta* **694**, pp. 227–277.

Specific Reading

Bevan, M.W. (1984). 'Binary *Agrobacterium* vectors for plant transformation', *Nucl. Acids Res.* **12**, pp. 8711–8721.
de la Pena, A. Lorz, H. and Schell, J. (1987). 'Transgenic rye plants obtained by injecting DNA into young floral tillers', *Nature* **325**, pp. 274–276.
Fromm, M., Taylor, L.P. and Walbot, V. (1986). 'Stable transformation of maize after gene transfer by electroporation', *Nature* **319**, pp. 791–793.
Grimsley, N., Hohn, T., Davies, J.W. and John, B. (1987). '*Agrobacterium*-mediated delivery of infectious maize streak virus into maize plants', *Nature* **325**, pp. 177–179.
Jones, H., Tempelaar, M.J. and Jones, M.G.K. (1987). 'Recent advances in plant electroporation', *Oxf. Surv. Plant Mol. Cell Biol.* **4**, pp. 347–357.
Klein, T.M., Wolf, E.D., Wu, R. and Sanford, J.C. (1987). 'High velocity microprojectiles for delivering nucleic acids into living cells', *Nature* **327**, pp. 70–73.
Reich, T.J., Iyer, V.N. and Miki, B.L. (1986). 'Efficient transformation of alfalfa protoplasts by intranuclear injection of Ti plasmids', *Bio/Technol.* **4**, pp. 1001–1004.
Senda, M., Takeda, J., Abe, S. and Nakamura, T. (1979). 'Induction of cell fusion of plant protoplasts by electrical stimulation', *Plant Cell Physiol.* **20**, pp. 1441–1443.
Tempelaar, M.J., Duyst, A., De Vlas, S.Y., Krol, G., Symons, C. and Jones, M.G.K. (1987). 'Modulation and direction of the electrofusion response in plant protoplasts', *Plant Science* **48**, pp. 99–105.
Tempelaar, M.J. and Jones, M.G.K. (1985). 'Fusion characteristics of plant protoplasts in electric fields'. *Planta* **165**, pp. 205–206.
Zimmermann, U. (1986). 'Electrical breakdown, electropermeabilization and electrofusion', *Rev. Physiol. Biochem. Pharmacol.* **105**, pp. 175–256.

Chapter 7

Manipulation of Plant Product Quality and Quantity

A major effort of conventional plant breeding programmes is directed towards the improvement of both the quality and the yield of specific products, whether they be carbohydrates, proteins, oils, fibre or secondary metabolites. Our knowledge of the molecular biological and biochemical basis of the regulation of the synthesis of these products is, in a number of cases, advanced enough for improvement by genetic manipulation to be attempted, and this is the subject of this chapter.

Photosynthesis

The most fundamental biochemical difference between plants and animals is the ability of plants to carry out photoautotrophic growth. During the 1920s and 1930s it was discovered that photosynthesis comprises two fundamentally different types of reaction, i.e. those which are dependent on the presence of light (the 'light' reactions) and those which are not (the 'dark' reactions). In the light reactions, the sun's radiant energy is absorbed by the chloroplastic pigment systems, and ultimately used to drive the phosphorylation of AMP and ADP and to generate reducing power, with the evolution of gaseous oxygen. The dark reactions involve the fixation of carbon dioxide and its subsequent integration into either three- or four-carbon organic acids. In the so-called C_3 plants, CO_2 is initially fixed by ribulose bisphosphate carboxylase–oxygenase (Rubisco) to form three-carbon phosphoglycerate and thereby enters the Calvin cycle (Fig. 7.1), whereas in C_4 plants, many of which are tropical species (and include a number of species of the Gramineae, Compositae, Chenopodiaceae, Euphorbiaceae, Amar-

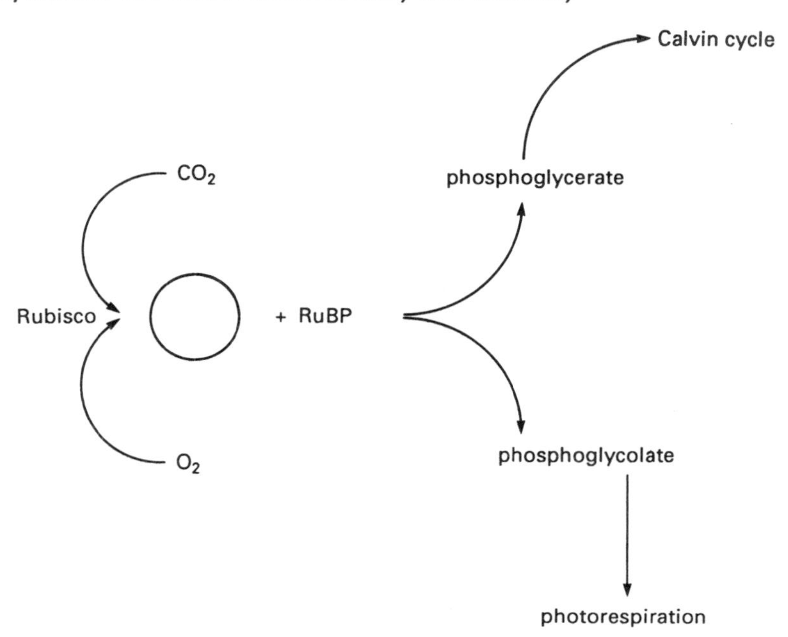

Fig. 7.1 The dual function of Rubisco. RuBP=Ribulose bisphosphate.

antheceae, Cyperaceae, Portulaceae, Nyctoginaceae and Zygophyllaceae), atmospheric CO_2 is fixed by phosphoenolpyruvate carboxylase (PEPCase) to form four-carbon oxaloacetate and then malate (Fig. 7.2). Male can be decarboxylated to release CO_2 to the Calvin cycle, which is localized in the bundle sheath cells (i.e. in a different cell type to PEPCase which is in the mesophyll tissue). This compartmental separation of the two CO_2-fixing enzymes is characteristic of C_4 plants.

C_3 and, to a lesser extent, C_4 plants may also utilize ribulose bisphosphate in an alternative set of reactions to the Calvin cycle, but still initially catalysed by Rubisco, to form glycolate (Fig. 7.1). This process requires oxygen and results in the release of CO_2, and is termed 'photorespiration'. For more details of the biochemistry of photosynthesis and photorespiration, the reader is referred to Hall *et al.* (1981).

Carbon fixation is the best studied area of plant metabolism, and for this reason, and also because of the great abundance of the key enzyme Rubisco, photosynthesis and photorespiration in particular have attracted much attention as processes amenable to manipulation by the techniques of genetic engineering. However, photosynthesis is obviously an integration of many complex processes, and before its successful manipulation can be achieved, much work is needed to understand the molecular biology and biochemistry of the system. For example, the activity of the Calvin cycle is also strongly influenced by the availability of

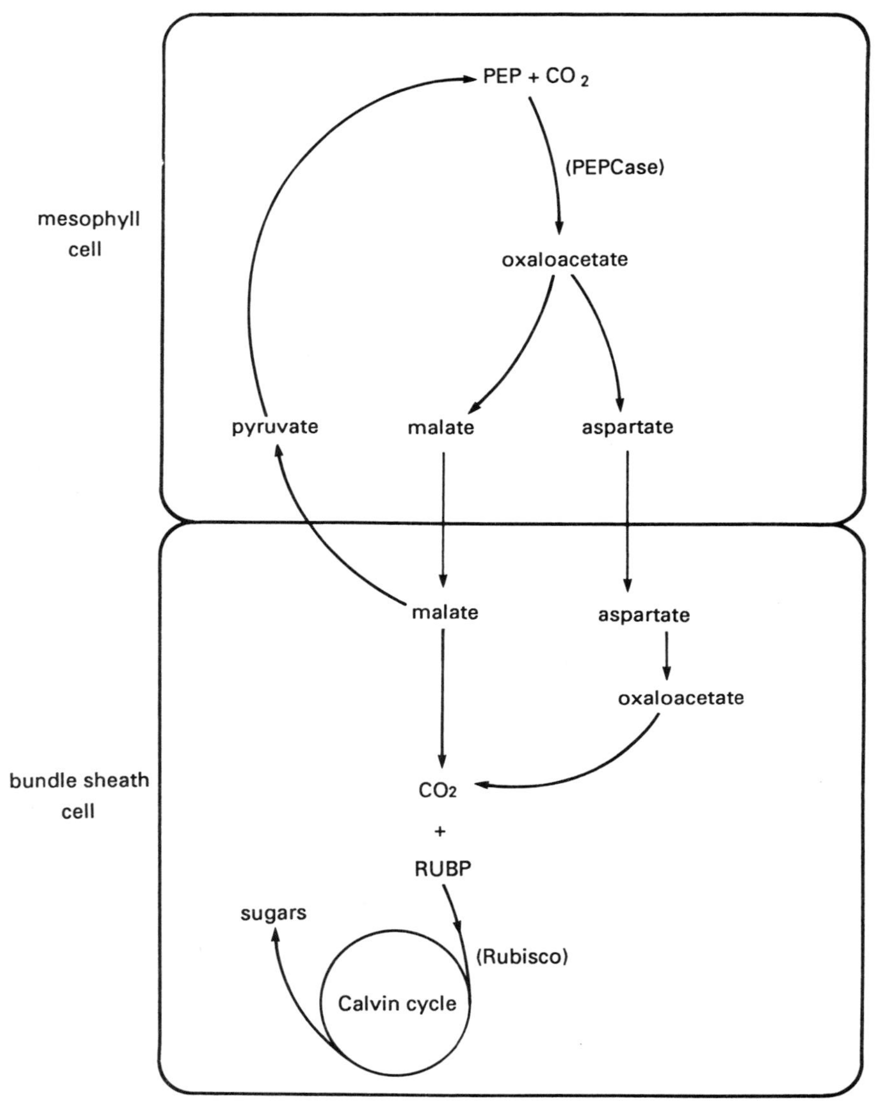

Fig. 7.2 Key features of C_4 metabolism. RuBP = Ribulose bisphosphate.

reducing power and phosphate energy, mediated by trans-membrane shuttle systems, which are potentially amenable to manipulation. The light-harvesting system of the chloroplast is similarly being actively studied. Finally, the distribution of the photosynthate within the plant, and also into root exudates, determines economic yield, and the manipulation of assimilate partitioning has obvious potential in increasing agricultural productivity. We will now consider in more detail some of the components of the photosynthetic system.

COMPONENTS OF THE PHOTOSYNTHETIC SYSTEM

Rubisco

The Rubisco molecule comprises eight identical large and eight identical small subunits, the genes of which are present in the chloroplast and nucleus, respectively. The genes coding for the small subunit constitute a multigene family, and were among the first plant nuclear genes to be isolated and characterized (Shinozaki *et al.*, 1983). Furthermore, DNA sequences, which determine light-inducibility, have been defined (review by Kuhlemeier *et al.*, 1987) and the entire chloroplast genome of tobacco has been sequenced. As we have seen, Rubisco is involved in two, apparently antagonist, processes (Fig. 7.1); firstly, by catalysing the fixation of carbon dioxide and hence providing carbon skeletons for intermediary metabolism; and, secondly, in the oxygenation of the same substrate, ribulose bisphosphate to form phosphoglycolate (photorespiration). In the latter process CO_2 is released and energy (ATP) is expended. Therefore, it may be possible to improve the yield of accumulated photosynthate either by increasing the relative rate of CO_2 fixation and/or by inhibiting the seemingly wasteful photorespiratory activity. It should be noted, however, that photorespiration may be essential for plant survival at low CO_2 concentrations, especially in high light intensity and/or high temperatures, and its complete elimination may be undesirable.

In order to attempt the genetic manipulation of either the carboxylase or oxygenase activities of Rubisco (for which there is a single active site), it is necessary to determine what factors regulate these processes *in vivo*, and whether they are amenable to such techniques. For example, both activities are dependent in part on the number of active enzyme molecules per cell and the affinity of the enzyme for substrate. It may be possible to increase the abundance of active molecules by stably introducing multiple copies of Rubisco genes into plants—in support of this possibility is the observation that there may be a direct relationship between neomycin phosphotransferase activity and the copy number of the encoding NPTII gene in transgenic plants or between alcohol dehydrogenase 1-S activity and copy number in transgenic maize callus; however, this relationship is by no means universal, and position effects may strongly influence the expression of introduced genes. It is also known that carboxylase activity is not simply a function of the number of Rubisco molecules present in the cell, but is modulated by the concentrations of its cofactors CO_2 and Mg^{2+}, and also by a range of organic molecules. A further potential difficulty is the targeting of multiple gene copies to specific organelles, but some significant advances have recently been made in this respect, with the discovery and manipulation of transit peptides (Chapters 3 and 9).

The carboxylase catalytic site is located within the large subunit of Rubisco, and it may be possible to raise carboxylase activity at a given CO_2 concentration, i.e. to increase the affinity for CO_2, by manipulating the structure of the large subunit itself, or of the small subunit, which may influence the structure of the catalytic site by inducing conformational changes in the large subunit. Such an alteration may also act to reduce photorespiration at the first step, since both CO_2 and O_2 are known to compete at the same catalytic site. It has been observed, for example, that when C_3 plants are grown in atmospheres enriched in CO_2, or depleted of O_2, increased yields of reduced carbon may be produced. Conversely, relatively high oxygen concentrations have been observed to be inhibitory to photosynthetic activity. In C_4 plants, Rubisco is located in the bundle sheath cells in which the CO_2 concentration is locally raised by the energy-dependent transport of four-carbon acids from the mesophyll cells (Fig. 7.2). The release of CO_2 in the bundle sheath cells acts to decrease the inhibitory effects of O_2 on photosynthetic assimilation, and in C_4 plants are characterized by low relative photorespiratory activity.

These observations have suggested the possibility of using protein engineering techniques to delete altogether the oxygenase activity of Rubisco, but whether this would be possible, while retaining carboxylase activity, is currently a matter for speculation. An alternative approach would be to select for mutant plants with a reduced oxygenase, relative to carboxylase, activity. It is known, in any case, that there is a natural variation between species in the ratio of the two activities, and studies of the molecular basis of such kinetic differences may lead to a genetic approach to modify the properties of Rubisco. Photorespiratory mutants of *Arabidopsis thaliana* and barley (Bright *et al.*, 1984) are known, but to date only mutations in the genes encoding enzymes in phosphoglycolate metabolism, but not Rubisco itself, have been produced. Such plants usually fail to survive when grown in natural air due to the accumulation of toxic intermediates of photorespiration, emphasizing the need for stable modification of the oxygenation reaction specifically (Keys, 1983, 1986).

Chloroplast Shuttle Systems

As indicated above, the activity of the Calvin cycle is influenced not merely by the abundance of Rubisco; and even if the energetically 'wasteful' processes of photorespiration could be decreased, other limitations to the system would become apparent. Apart from CO_2 availability, the availability of phosphate energy and reducing power are critical to ensure photosynthetic activity. Enzyme localization studies indicate that, in leaves, sucrose cannot be synthesized in the chloroplast and is unable in any case to cross the chloroplast envelope. Most of the phosphate energy required for sucrose synthesis comes from the chloroplastic light reaction, and while the translocation of molecules of ATP, ADP and AMP may account for a small proportion of the energy transfer needed (due to the low capacity of the translocator), indirect transport systems appear to function as high capacity translocators of phosphate. The most efficient of such systems appears to be a phosphate shuttle in which 3-phosphoglycerate and dihydroxyacetone phosphate are exchanged between the stroma and cytoplasm in a specific import–export mechanism (Fig. 7.3). The inward movement of 3-phosphoglycerate also supplies some phosphate for the chloroplastic synthesis of ATP, triose phosphates and pentose. Reducing power is also exported from the chloroplast as dihydroxyacetone phosphate, and also via malate/oxaloacetate (dicarboxylate) shuttles (Fig. 7.4); the chloroplast envelope does not allow the direct transport of pyridine nucleotides. The function of the carrier molecules, presumed to be membrane-located proteins, is extremely specific, but their structures have not

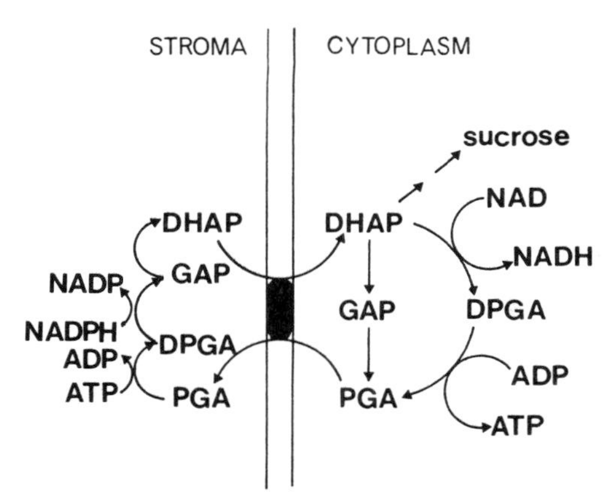

Fig. 7.3 Proposed chloroplast phosphate shuttle. This shuttle works in the light to transfer phosphate and reducing power, in the ratio of H : ATP = 2 : 1, into the cytoplasm. In the dark, the shuttle works in the opposite direction. PGA = 3-phosphoglycerate; GAP = glyceraldehyde 3-phosphate; DPGA = 1,3-diphosphoglycerate; DHAP = dihydroxyacetone phosphate.

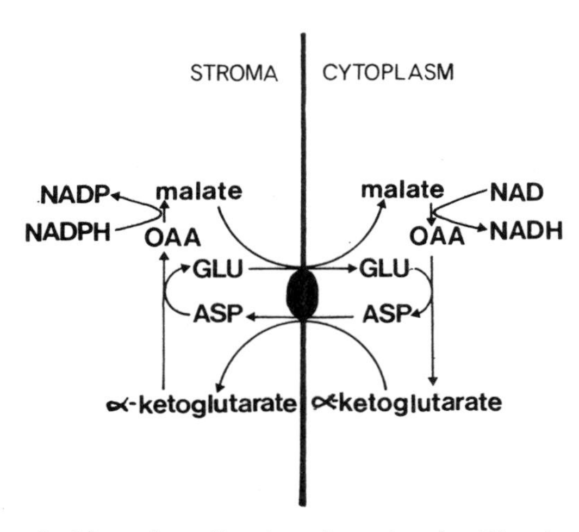

Fig. 7.4 Proposed chloroplast dicarboxylate shuttle. Illuminated chloroplasts reduce oxaloacetate to malate, the transport of which across the membrane controls the export of reducing power. The inward movement of oxaloacetate is probably linked *in vivo* to the glutamate – oxaloacetate transamination reaction, the product of which is α-ketoglutarate. OAA = oxaloacetate; GLU = glutamate; ASP = aspartate.

been resolved, and the genes have not been isolated. The gene for the mitochondrial adenine nucleotide translocator protein of maize has, however, been cloned and sequenced. It should be apparent that, as modifications to Rubisco activity are achieved, other limitations to Calvin cycle activity will arise, and manipulation of phosphate energy and reducing power availability may become important. It is conceivable that, with greater knowledge and understanding of the shuttle systems, this could be achieved by genetic engineering of the carrier molecules themselves.

Light-Harvesting System

The efficient transfer of light energy, absorbed by chlorophyll to the photosynthetic reaction centres, is dependent upon the light harvesting pigment proteins located in the thylakoid membrane. They are aggregated into two types of protein complex, designated light-harvesting complex-I (LHC-I) and II (LHC-II), which sensitize the reaction centres of photosystems I and II, respectively. LHC-I and -II each contain unique complements of polypeptides synthesized in the cytoplasm as precursors (apoproteins) and possessing, at their N-terminal ends, transit peptides which are post-translationally cleaved during transport into the chloroplast. Most detailed studies have dealt with LHC-II (also known as the chlorophyll *a*/*b* binding protein), which is extremely abundant. Like the small subunit of Rubisco, the apoproteins of LHC-II are encoded by multiple nuclear genes, and in *Petunia*, 16 genes, comprising five multigene families, have been identified. The number of different polypeptides which complex to form LHC-II is, however, still not known with certainty.

There are three other categories of protein complexes located in the thylakoid membrane: the reaction centre polypeptides, which determine the relationship between the chlorophylls and the electron donors and acceptors: i.e. at photosystems-I and -II (PS-I and PS-II); apoproteins of electron transport carriers, which bind the various cofactors (e.g. haem groups) involved in the transport of electrons along and across the membrane to generate the chemiosmotic potential; and, finally, the polypeptides of the 'CF_1/CF_0' coupling factor, the complex which links the energetic transmembrane proton gradient to drive the photophosphorylation of ADP. For a review of each of these complexes, the reader is referred to Timko *et al.* (1985). A number of the genes for component polypeptides have been identified and sequenced, and some are listed in Table 7.1; references to the original work are cited in Timko *et al.* (1985). Of these genes, several are light-regulated, and together with the light-regulated, but nuclear-encoded genes for LHC-II and the small subunit of Rubisco, constitute the 'photogenes', all of which appear to be under the control of phytochrome. The level at which phytochrome acts, however, is still not fully understood, but further work on the structure and regulation of both photogenes and phytochrome itself (see Chapter 8) may indicate strategies by which the photosynthetic process may be manipulated.

One approach to investigate the functional significance of specific DNA sequences of photogenes has been to link them to marker genes, such as CAT. For example, Timko *et al.* (1985) made a series of Bal-13 deletions of the 5′ region of a

Table 7.1 Some plastid-encoded genes of thylakoid membrane proteins which have been sequenced

• Core components of PS-II	• Components of Cyt b_6f complex
32-kDa Q_B protein	Cyt *f*
D_2 polypeptide	Cyt b_6
47-kDa RC apoprotein	subunit 4
43-kDa RC apoprotein	• Components of the CF_1/CF_0 complex
Cyt b_{559}	β-subunit of CF_1
• PSI RC apoprotein A_1	ε-subunit of CF_1
• PSI RC apoprotein A_2	α-subunit of CF_1
	subunit-I of CF_0
	subunit-III of CF_0

PSI/II = Photosystem I/II; RC = Reaction centre

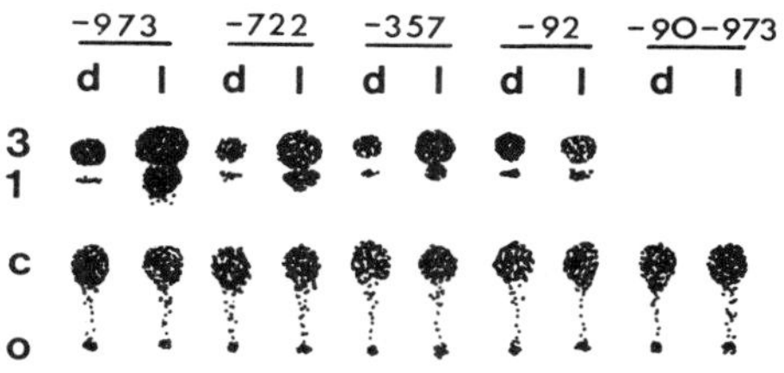

Fig. 7.5 Inducibility of the pea Rubisco small subunit gene promoter by light. *Bal*-31 deletions were made of the 973-base-pair 5′ flanking region, which were linked respectively to the coding region of the CAT marker gene. Expression was measured as CAT activity in transformed tobacco callus in light (l) or dark (d). The length of the 5′ region remaining after deletion (−973, −722, −357, −92) is indicated for each transformant. −90−973 represents the deletion of nucleotide sequences, including the TATA box, from the 973-base-pair promoter fragment (4–90 base pairs upstream of the CAP site). *CAT assay*: O = origin of TLC spots; C = chloramphenicol (substrate for CAT enzyme); 1 and 3 = acetylated reaction products 1-acetylchloramphenicol and 3-acetylchloramphenicol, respectively. The size of these spots is related to the activity of the CAT marker enzyme. (After data from Timko *et al.*, 1985.)

chimaeric gene comprising the Rubisco small subunit promoter linked to the CAT coding region. These truncated genes were introduced into tobacco cells by *Agrobacterium*-mediated transformation, and the transformed callus, which was cultured in either the light or the dark, was subsequently assayed for CAT enzyme activity (Fig. 7.5). It was found that at least 722 base pairs upstream from the CAP site were required for high levels of light-inducible gene expression: as the deleted regions were increased in size, to produce a 5′ region of 350 base pairs upstream, promoter activity was concomitantly reduced. Even with 973 base pairs upstream to the coding region, dark grown callus expressed only very low levels of CAT activity. This type of approach is invaluable for dissecting genes into their functional components, a prerequisite for the manipulation of regulation.

PARTITIONING OF PHOTOSYNTHETIC ASSIMILATE

The most important factor in determining economic yield is not necessarily the total capacity of a crop for photosynthesis, but the way in which assimilates are distributed within the plant, either for continued vegetative growth or for accumulation in storage organs, usually roots, tubers, seeds or fruit. The so-called source–sink relationships are extremely complex. It is often unclear how the rate of assimilate accumulation ('sink strength') is limited by the production of photosynthate by the rate of translocation of assimilate or uptake and deposition by the storage organ, or the extent to which sink size can influence photosynthetic rate. For the optimum product yield, it is essential that the expenditure of photosynthate on other tissues should be the minimum required to produce a plant habit which results in the efficient arrangement of the leaves to optimize the capture and conversion of solar energy. Assimilate is also lost through the roots as polysaccharide exudates, a phenomenon strongly influenced by water availability and the composition of the microflora in the rhizosphere. Labelling studies indicate that between 6–15% or more of all carbon fixed by a plant may be lost in this way.

The achievement of a balance in the development of 'producing' and 'accumulating' organs, and of the improved partitioning of photosynthate assimilates, have been primary aims of plant breeding programmes for many years. Since the physiological and genetic basis of these developmental phenomena are largely unknown, the prospects for their genetic manipulation are currently difficult to define. It is most likely that the proportion of source and sink organs, in morphological terms, is determined polygenically, and RFLP analysis may prove useful, in the long term, in identifying the genes involved. Similarly, there is undoubtedly a genetic element in the determination of assimilate accumulation, which is also likely to be under polygenic control. Obviously, much more biochemical and physiological research must be carried out before significant advances in this area are made.

Nitrogen Fixation

Atmospheric nitrogen is extremely inert. Despite its great natural abundance i.e., comprising almost 80% of the air, it is in a form which is not directly available to plants or animals, but must be acquired as a component of other molecules. The biochemical capability to 'fix' nitrogen, i.e. to separate nitrogen atoms and recombine them with hydrogen or oxygen, is limited to prokaryotic organisms such as the aerobic (e.g. *Azotobacter*) and anaerobic (e.g. *Klebsiella*) bacteria and the cyanobacteria (e.g. *Anabaena*, *Nostoc*, formally described as 'blue-green algae').

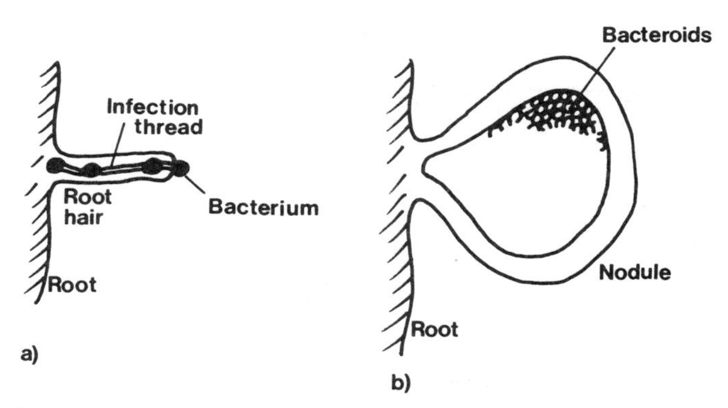

Fig. 7.6 Diagrammatic representation of root nodule formation. (a) Infection by the *Rhizobium* bacteria takes place via the root hairs by the invagination of the cell wall to form an 'infection thread'. (b) Release of the bacteria from the infection thread leads to the formation of the nodule, packed with nitrogen-fixing bacteroids.

However, some termites and shipworms, and certain genera of plants, notably most members of the Leguminosae but also the aquatic fern *Azolla* and tropical grasses such as *Digitaria*, are able to develop symbiotic relationships of some form with nitrogen-fixing microorganisms, and thereby flourish in nitrogen-poor environments (Stewart and Gallon, 1980). Many legumes, with which we are especially concerned here, are characterized by the presence of 'root nodules', swellings on the roots which harbour *Rhizobium* bacteria (Fig. 7.6). The molecular biology of this association is one of intense interest, both from the point of view of studying the plant–bacteria interactions (i.e. recognition, coordination of gene expression, coordination of metabolism) and also because of the potential for the genetic manipulation of this important biochemical process.

The nitrogenase system of bacteria catalyses the reduction of dinitrogen (N_2) to ammonia in an energy-demanding process (Fig. 7.7). The overall reaction is the same as that in the industrial Haber–Bosch process, with the breaking of the dinitrogen triple bond, followed by the addition of three hydrogen atoms to each nitrogen. In the industrial process, hydrogen is supplied in gaseous form, while in the cell it is derived from the oxidation of carbohydrates, via an electron-transport chain. The nitrogenase enzyme comprises two polypeptides, designated Components I and II, respectively. Thus, electrons flow from a reduced carrier molecule, such as ferredoxin or flavodoxin to Component II (also known as nitrogenase), which then reduces Component I (dinitrogen reductase) which in turn catalyses the reduction of N_2 to NH_4^+. No free intermediates between nitrogen and ammonia have been detected.

When purified nitrogenase is incubated with ATP in the *absence* of nitrogen, hydrogen gas is evolved. This secondary activity of nitrogenase is wasteful in energetic terms; and hydrogen itself is a competitive inhibitor of nitrogenase

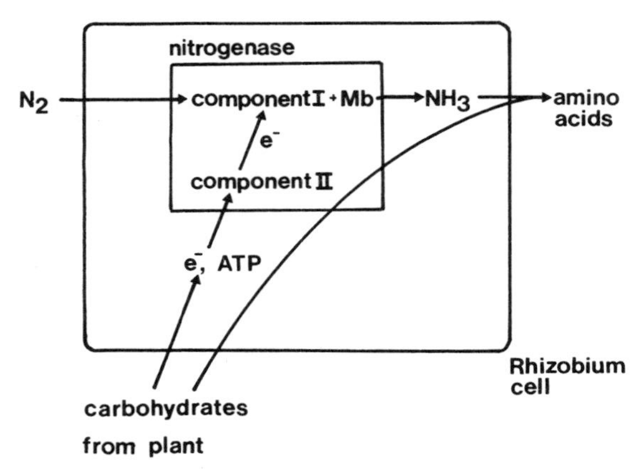

Fig. 7.7 Nitrogen fixation in *Rhizobium*. Mb = molybdenum.

activity. Nitrogenase is also extremely sensitive to an inhibitory effect of oxygen (Witty *et al.*, 1986), and this susceptibility may in part explain the limited species distribution of nitrogen fixation.

The proteins involved in fixation are encoded by a number of *nif* ('nitrogen fixing') genes (e.g. at least 15 for *Klebsiella pneumoniae*), clustered either on the bacterial chromosome or, as is the case for some *Rhizobium* species, on large extra-chromosomal plasmids. The symbiotic association between root and *Rhizobium* is determined by a further set of bacterial genes, the *nod* ('nodulation') genes, and also by plant genes which encode the 'nodulin' proteins, a group of at least 30 nodule-specific polypeptides. The developmental regulation of some of the nodulin genes has been studied (e.g. Campos *et al.*, 1987). The infection process comprises a series of distinct stages: recognition of the host root; curling of the root hair, which is the site of entry of the bacteria: the invagination of the cell wall of the root hair to form an 'infection thread', through which the bacteria are able to reach the cells of the root cortex; and, finally, the release of the bacteria into the cortical cells, which become enlarged and differentiate into the nodule. The bacteria themselves develop into nitrogen-fixing 'bacteroids'. Specific legumes are associated with a particular species of *Rhizobium*: alfalfa is infected by *R. meliloti*, soybean by *R. japonicum* and clover by *R. trifolii*. This specificity appears to be determined by protein recognition factors on the surfaces of root hairs, which bind to sugar receptor sites on the bacterial cell surface.

GENETIC ENGINEERING OF NITROGEN FIXATION

There are at least four generalized strategies which can be adopted to improve the nitrogen contents of plants grown on poor soils by genetic engineering of nitrogen fixation. These will now be discussed.

Increasing the Host Range of Symbiotic Nitrogen Fixing Bacteria

The genetic determinants of nodulation, as indicated above, are located in both plant and bacterial cells. A first step in increasing the range of symbiosis is the relatively easy manipulation of the bacterial genes. Plasmid-borne *nod* genes could be transferred from nodulating to non-nodulating strains of nitrogen-fixing bacteria. In relation to this, it can be demonstrated that if a plasmid from *Rhizobium leguminosarum*, which normally forms nodules on peas, is transferred to *R. phaseoli*, which normally forms nodules only on bean plants, the ability to nodulate peas is also transferred. The role of the host plant in determining the success of the symbiotic relationship is less easy to define, and has been investigated primarily by the characterization of nodule-specific gene products, the nodulins. Arguably the most important, and certainly the most abundant of these proteins, is leghaemoglobin, representing 20–30% of the total protein content of the nodule. It is a product of a multigene family and is structurally and functionally related to the animal proteins haemoglobin and myoglobin, having a high affinity for gaseous oxygen. Because of the strong inhibition of nitrogenase activity by oxygen, it is believed the leghaemoglobin maintains a stable O_2 concentration within the nodule by acting as an oxygen carrier. All nodulins (including leghaemoglobin) can be categorized into three groups:

Group 1 nodulins represent structural proteins of the nodule, but specific functions have yet to be assigned to them. Due to the association of a number of the polypeptides with nodule membranes, it has been suggested that some Group 1 nodulins may play a role in the transport of nodule metabolites.

Group 2 nodulins include enzymes involved in the further metabolism of fixed nitrogen, such as glutamine synthase and uricase. Glutamine synthase catalyses the conversion of ammonia and glutamate to glutamine, which is in turn metabolized to other amino acids and ultimately protein. Uricase is present in nodule cells lacking bacteroids, and mediates the catabolism of uric acid, a breakdown product of purines, to allantoin which, with allantoic acid, appear to be major forms of stored or transported nitrogen.

Group 3 nodulins are described as those proteins which support bacteroid function, and include leghaemoglobin. The complexities of the plant-encoded aspects of the symbiosis are therefore apparent, and it would be expected that any modification of this relationship, such as the extension of host range, would require the transfer of these, as well as the relevant bacterial, characters.

Improving the Efficiency of Nitrogen Fixation

It is conceivable that, while our knowledge of the mechanisms of nitrogen fixation is far from complete, the efficiency of the process may be improved. One approach might be to screen different strains of *Rhizobium* for a high capacity for fixation in association with a particular host species. Mutagenesis would presumably generate metabolic diversity, but this approach is rather empirical, with problems of selecting desirable strains. Similarly, of the 10 000 or so species of the Leguminosae, only a relatively small number (approximately 10%) have been studied for their ability to enter nitrogen-fixing symbiosis. It seems likely that many other, perhaps exotic, species would develop high nitrogen-yielding associations.

An alternative approach to improving the efficiency of fixation would be to attenuate the energetically wasteful evolution of hydrogen gas by nitrogenase. In relation to this, it is known that alder (*Alnus rubra*, a non-legume) and cowpea, which do form root nodules, do not release gaseous hydrogen; the gas is thought to be recycled in the cell. If the molecular basis for this energy recycling was determined, it would be valuable to attempt to transfer the character to hydrogen-evolving species. Indeed, a gene encoding a H_2-uptake hydrogenase has been identified which is able to recapture evolved hydrogen. The gene has been transferred to various strains of *Rhizobium*, with the result that, when in association with legumes, more efficient nitrogen fixation and improved plant growth was achieved.

Nitrogen Fixation in the Presence of Nitrate

Nitrogen fixation is suppressed in the presence of environmental fixed nitrogen, due to the inhibition of nitrogenase synthesis. Nevertheless, it has been possible to isolate mutant strains of *Azotobacter* which fail to exhibit this inhibitory regulation and will even excrete ammonia into the extracellular environment. Such strains could perhaps be used to inoculate the soil of fields of crops unable to fix their own nitrogen in an artificial symbiosis; or they could be cultured in bioreactors to produce fixed nitrogen under controlled environmental conditions, a process which could become increasingly important as natural nitrate deposits become depleted.

Transfer of Nitrogen-fixing Genes to Non-symbiotic Plants

It will be apparent from previous chapters that significant advances have now been made in the techniques for transferring specific genes from bacteria to plants, and in the manipulation of the expression of those genes. It may therefore be possible to bypass the symbiotic association altogether and introduce the nitrogen-fixing genes directly into non-leguminous species. The clustered genes have been successfully transferred from *Klebsiella pneumoniae* to *E. coli* and nitrogenase activity was detectable if the transformants were protected from inhibitory oxygen. This latter observation illustrates the point that the engineering of a nitrogen-fixing plant is a formidable target, requiring the introduction not only of the bacterial *nif* genes, but presumably also at least some of the plant nodulins. It is thus very obvious that, before there is ever any hope of this being achieved, much more basic characterization of the system must be carried out.

Recently, it has been demonstrated that novel, nitrogen-fixing actinomycete fungi can be produced by fusing protoplasts of non-fixing *Streptomyces* with nitrogen-fixing *Frankia*, which is normally associated symbiotically with alder. Regenerated hybrids were capable not only of fixing nitrogen but also of developing a symbiotic relationship with alder, a natural host species of *Frankia* (Prakesh and Cummings, 1988). This demonstrates the possibility of transferring not only nitrogen-fixation genes and determinants for symbiosis, but also the genes responsible for the protection of nitrogenase against oxygen.

Solute Uptake

The establishment of plants in a particular habitat is strongly influenced by the ionic composition of the soil. We have seen how some species have evaded the problems of low nitrogen soils by entering symbiotic relationships; others may adopt a carnivorous lifestyle by entrapping insects in specially modified leaf structures. It has been recognized for many years that the growth and development of crop species can be improved by the application to the soil, not only of nitrogen but also of phosphate, potassium, sulphur and microelements. The question then arises whether the crop yield is limited by the ability of the plant to recover these ions from the soil, either because of their chemical sequestration, or because of a limited availability of transmembrane carrier sites. A different symbiosis to that described above is the association of mycorrhizal fungi with the roots of many species, and the beneficial effects for the plant in terms of improved nutrient uptake is well characterized. Furthermore, soil bacteria may also aid mineral ion availability, for example, by solubilizing phosphate. The molecular bases of recognition systems involved in fungal or bacterial symbiosis are, however, not known, and it appears that almost any plant can develop a mycorrhizal association to a greater or lesser extent. Therefore, the attention of the molecular biologist is drawn to the study of ion-transport mechanisms. Due to the complexity of such mechanisms, the direct manipulation of solute uptake is likely to be many years away; however, some molecular information is available, particularly in relation to membrane-bound ATPases. Early work with animal systems, such as the membranes of kidney cells and erythrocytes, and with isolated mitochondria and chloroplasts, demonstrated that the transport of K^+ and Na^+ and of H^+ was linked to the activity of ATPases. It was observed, for example, that when erythrocyte ghosts (i.e. hypotonically-stressed red blood cells) were loaded with K^+ ions and placed in a Na^+ solution, K^+ ions moved down a diffusion gradient and Na^+ ions moved in; this ion exchange was able to drive ATP synthesis. What is the evidence that ATPases may be involved in solute uptake in plants? A hypothetical model mechanism is illustrated in Fig. 7.8, in which ion uptake is driven by ATP. Such a model requires certain criteria to be fulfilled: are ATPases localized at membranes?; is there a membrane potential and pH gradient across the membrane?; and can ATPase activity and the ion gradients account for observed rates of ion uptake? ATPases have been shown by histochemical studies to be located on the plasma membrane of plant cells by histochemical studies. Lead nitrate is added to tissue sections, and, if the active enzyme is present, is converted to lead phosphate using inorganic phosphate derived from ATP; on exposure to hydrogen sulphide gas, the lead phosphate is precipitated as black lead sulphide. Physiological assays have exploited the fact that K^+ ions stimulate ATPase activity, and this stimulation has been shown to occur in isolated plasma-membrane fractions. Plant cells do have membrane potentials, which can be reduced by the action of pore-forming antibiotics such as nystatin, but present evidence suggests the ATPase activity alone cannot account for observed rates of ion uptake. Electron microscopical evidence indicates that ions may be taken up by the invagination of the plasma membrane to form ion-carrying vesicles, and the uptake activity of a plant may be influenced by factors

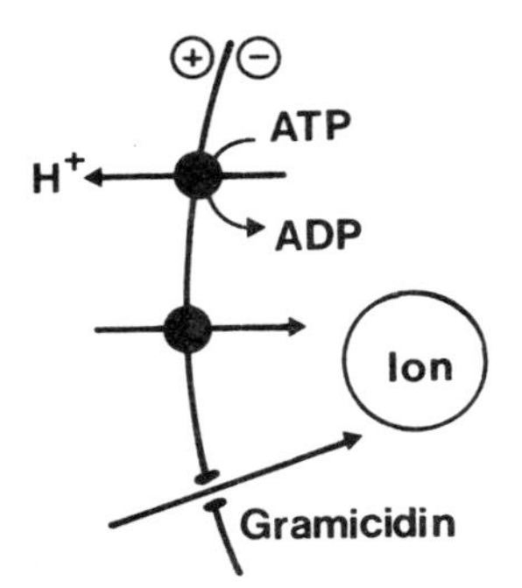

Fig. 7.8 One model for ion uptake driven by ATP. Here, a transmembrane proton gradient is established, powered by the hydrolysis of ATP. This gradient in turn provides the energy for the uptake of ions, via specific carriers. Pore-forming antibiotics such as gramicidin should allow the uptake of ions (e.g. K^+) down an electrical gradient; this is observed.

such as the rates of transpiration, photosynthesis and plant growth (and, without doubt, interactions between these factors). It is difficult to relate such parameters with molecular mechanisms of uptake, but it can be hypothesized that, under a range of environmental factors or of stages of plant development which are observed to influence the uptake of specific ions, specific carriers may be switched on or off. Very little is currently known about such regulatory mechanisms, but studies using existing uptake mutants (such as nitrate uptake mutants of barley) should prove valuable.

Nutritional Quality

While the yield of a crop is easy to define, the definition of nutritional quality is a more complex matter, and may vary according to what is considered 'healthy'. For example, the ratio of saturated to unsaturated lipids in oil crops is of interest in relation to problems of heart disease, and this area will be considered later on. But because of the great economic and nutritional importance of cereals and legumes and because protein quality is limiting in many of these species, much effort has been invested in the study of the biochemistry and molecular biology of seed development, and particularly of seed storage proteins.

PROTEIN CONTENT

Applied interests relate to the fact that the storage proteins of most major species are deficient in one or more of the amino acids essential for animal growth. For example, wheat, barley, maize and *Sorghum* accumulate major storage proteins which are low in lysine (Bright and Shewry, 1983), while storage proteins of legumes are deficient in sulphur-containing amino acids. Barley and *Sorghum* are also low in threonine, and maize in tryptophan. Therefore, manipulation of the nutritional quality of proteins involves an alteration of the total amino acid complement of the seed to provide a nutritionally-balanced product.

Genetic engineering techniques can provide at least two strategies to improve protein quality. The first is by increasing the proportion of a specific amino acid within a protein. This approach would require the isolation and cloning of the gene for the protein, and modifying its primary structure by chemically inserting additional codons for the limiting amino acids. The modified gene, along with its regulatory sequences, would then be reintroduced into the species. Difficulties with this approach include a lack of reproducible techniques for stable cereal transformation (although methods have been developed for a number of legume species), and problems associated with alterations in the physico-chemical properties of proteins, and perhaps as a corollary an altered stability in the plant as a result of an altered primary structure. Furthermore, the major storage proteins are encoded by multigene families. Nevertheless, the genes of a number of plant storage proteins have been cloned, and expressed in transgenic plants of the same and different species as the original; included are, for example, the zein genes of maize, the B_1 hordein gene of barley, the glutenin genes of wheat and the glycinin and conglycinin genes of soybean (see Chapter 9).

A second approach to the improvement of the nutritional quality of protein is to modify the expression of existing genes, so that proteins which are relatively rich in the deficient amino acids are preferentially synthesized. This strategy requires detailed analysis of the structure and regulation of specific genes, and much progress has been made in this respect. For example, sequences essential for the regulation of the tissue-specific transcription of storage proteins have now been identified by ligating them to reporter genes. Thus, sequences directing tuber- and root-specific expression of the potato storage protein patatin have been isolated and linked to both CAT and β-glucuronidase (GUS) (Twell and Ooms, 1987), and upstream elements have been shown to regulate the seed-specific expression of various seed storage genes (see Chapter 9). These are the first steps towards the direct genetic manipulation of storage protein composition.

A different approach involves the generation and selection of mutants. A mutant can be defined as a cell, cell line or organism in which the primary sequence of the DNA has been permanently altered. Phenotypically variant plants which may arise through somaclonal variation can only be described as mutants if molecular or genetic evidence demonstrates one or more irreversible changes to the chemical structure of the genome. Mutants can be generated by both chemical and physical mutagens (see Chapter 6), and selection can be carried out at all levels of tissue complexity: from protoplasts and single cells to somatic embryos, seeds and even intact whole plants. Mutant genes which influence the synthesis and accumulation of both starch and the alcohol-soluble storage proteins ('prolamins') have been described in the diploid cereals maize, barley, *Sorghum* and rice, and are commonly selected on the basis of an altered endosperm phenotype, such as the *shrunken* and *waxy* mutants of maize. Direct screening of the world barley collection for lines with increased levels of basic amino acids has lead to the identification of a 'high lysine' genotype designated 'Hiproly' because of its increased contents of protein and lysine. Hiproly contains approximately 30% more lysine than other cultivars, but has a lower yield due to smaller seed size. The high lysine character is controlled by a single recessive gene, *lys*, and specific proteins containing a high proportion of lysine have been identified. Four such proteins are β-amylase (5.0% w/w lysine), protein Z (7.1% w/w lysine), chymotrypsin inhibitor CI-1 (9.5% w/w lysine) and chymotrypsin inhibitor CI-2 (11.5% w/w lysine). cDNA clones for the chymotrypsin inhibitors CI-1 and CI-2 have been made from poly(A^+) RNA isolated from Hiproly endosperm, and the transfer of the high lysine character to other genotypes is now a possibility. Other high lysine mutants have also been generated, using both physical and chemical mutagens (see Shewry *et al.*, 1987). Unfortunately, however, all, including Hiproly, have a reduced starch content and hence reduced total yield.

It is also possible to use *in vitro* techniques to produce mutant lines which over-produce specific amino acids by selecting against amino acid analogues. The analogues are toxic either because they are incorporated into protein or because they falsely feedback-inhibit the synthesis of the natural amino acid. Death will result if the cell is unable to synthesize unusually high endogenous levels of the natural amino acid, for example, if a key enzyme is amplified in activity due to multiple gene copies or to insensitivity of the enzyme to negative feedback regulation. It is not always necessary to mutagenize cultured cells to generate resistance, for the genetic variation which 'naturally' exists within a population of suspension culture cells may be adequate to select cells carrying a specific mutation. Again, however, it is emphasized that molecular evidence is required to designate such a cell line as a mutant, since resistance to toxic analogues may be due, not merely to mutant enzymes with altered kinetic properties, but also to gene amplification or to reduced activity of an uptake mechanism.

One example which has been well characterized is the over-production of tryptophan by cultured cells of tobacco and carrot in the presence of the tryptophan analogue 5-methyltryptophan (5MT) (Gonzales and Widholm, 1985). When suspension culture cells were inoculated in a medium containing inhibitory levels of the analogue, but without a mutagen, resistant colonies appeared at a frequency of 10^{-6} to 10^{-7}. Mutagenesis of carrot cultures increased this frequency 10–100-fold, and enzymological studies demonstrated altered properties of anthranilate synthase, the first enzyme in that part of the aromatic amino acid pathway leading to tryptophan (Fig. 7.9). This enzyme was demonstrated to be less sensitive to negative feedback by both 5MT and to tryptophan, resulting in the over-production of free tryptophan and hence survival of the cell line.

Both potato and tobacco cells, while resistant to 5MT in liquid suspension, failed to pass on this phenotype to plants which were regenerated. It was discovered that cultures of these species possessed two forms of anthranilate synthase, one of which is sensitive to 5MT and one of which is resistant. Wild-type cell lines contain a predominance of the sensitive form, while in resistant cell lines the resistant isoenzyme is abundant. Regenerants from resistant cells, however, expressed primarily the sensitive form. In contrast to this it was found that both carrot and *Datura innoxia* possess only one form of the enzyme, and plants which

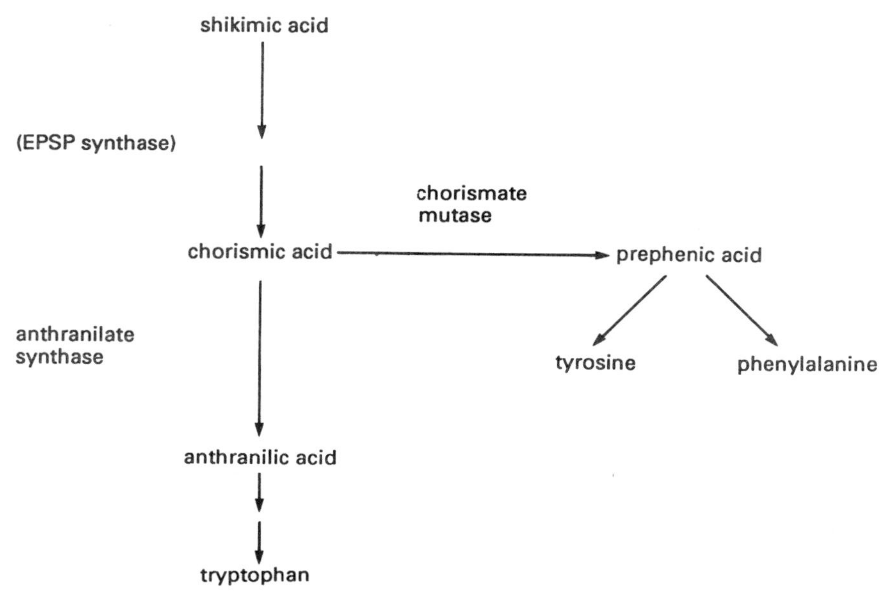

Fig. 7.9 Summary of the biosynthesis of aromatic amino acids. EPSP synthase = 5-enolpyruvylshikimate-3-phosphate synthase.

have been regenerated from 5MT-resistant cells not only also express resistance but also pass it on to their progeny.

The selection of amino acid over-producers is therefore of potential in the modification of nutritional quality and also for the production of specific, valuable secondary metabolites derived from amino acids (see later). On the other hand, the use of amino acid analogues as herbicides has attracted interest, and this is considered in Chapter 9.

LIPID CONTENT

The most abundant forms of lipids in plants are the triacylglycerols (triglycerides), which are accumulated as storage products in seeds or in certain fruits (e.g. avocado) and are a commercially important source of fats and oils. The physical and chemical properties of fats are dependent upon the fatty acid composition of the lipids and the distribution of the fatty acids on the glycerol skeleton. Fatty acids are classified as being either saturated, possessing no double bonds in the carbon chain, such as lauric (12 : 0, where '12' represents the number of carbons and '0' represents the number of double bonds), myristic (14 : 0), palmitic (16 : 0) and stearic (18 : 0) acids, or as unsaturated, with one or more double bonds, for example oleic (18 : 1), linoleic (18 : 2) and linolenic (18 : 3) acids. These seven fatty acids account for about 94% of the total fatty acids of commercial oils, but the precise composition of storage triacylglycerols varies between species and, in a particular species, varies with the developmental age of the storage organ and between different tissues within an organ. Mammals are unable to desaturate oleic to linoleic acid, and so this fatty acid, which is an essential dietary component, must be obtained from plants. Although the major fat in lipid-rich tissues is triacylglycerol, there are exceptions such as jojoba oil, which is composed of liquid wax esters, and *Raphia* palms, which accumulate high levels of polar lipids (glycolipids and phospholipids). Pea seeds are poor in triacylglycerols, where glycolipids and phospholipids are the predominant fat. Tissues which accumulate more than approximately 10% on a dry weight basis of fats are described as 'lipid-rich', and this is usually due to high levels of triacylglycerols.

During the development of lipid-rich seeds (discussed in more detail in Chapter 8), there is generally a lag phase of 10–30 days after pollination during which time there is little accumulation of fat. This is followed by a phase of accumulation for 2–5 weeks, which is in turn followed by a period of little change in lipid content. During the accumulation phase there are dramatic changes in fatty acid composition, reflecting decreasing relative levels of polar lipids and increasing levels of triacylglycerols.

What are the prospects for manipulating triacyglycerol production in storage tissues? A number of questions can be asked, such as, why do peas produce only very low levels of triacylglycerols?; what regulates the onset of triacylglycerol biosynthesis?; and what controls the fatty acid composition of the accumulated lipids? The biochemistry of lipid metabolism is very complex, with more than 30 enzymes involved in the pathway from acetyl-CoA to accumulated product. The production of acetyl-CoA from carbohydrates can be considered as the first committed step of fatty acid biosynthesis in plants, since, once it is made in the chloroplast, it appears not to be available for the TCA cycle or other synthetic reactions. The developmental and tissue-specific regulation of the synthesis of specific fatty acids in under very tight control—for example, a specific genotype of safflower (*Carthamus tinctorius*) produces a fixed proportion of oleic and linoleic acids in the seed oil, with no detectable linolenic acid, while the leaves, on the other hand, contain a high proportion of linolenic acid. Some information is available on the genetic regulation of fat synthesis. Traditional plant breeding techniques have essentially eliminated long-chain fatty acids such as erucic acid (22:1) in the seeds of oil seed rape (*Brassica napus*), but the biochemical basis of this is unclear. Genetic analysis suggests that the erucic acid content is controlled by two genes which display no dominance and act in an additive fashion. The regulatory effects of these genes has been determined to act on the chain-lengthening pathway from oleic acid to eicosenoic (20 : 1) and erucic acids, presumably either at the chain-lengthening steps themselves or at the steps regulating the supply of malonyl-CoA, which is essential for the synthesis of 22-carbon erucyl-CoA and thence erucic acid from 20-carbon eicosenoyl-CoA (Fig. 7.10). By reducing the biosynthesis of erucic acid, the precursor oleic acid accumulated to maintain a constant oil content in the seed.

Obviously much more work must be done to identify and isolate key regulatory enzymes. One approach would be to identify tissue- or developmental-specific transcripts and make cDNAs to isolate genes which are switched on during the

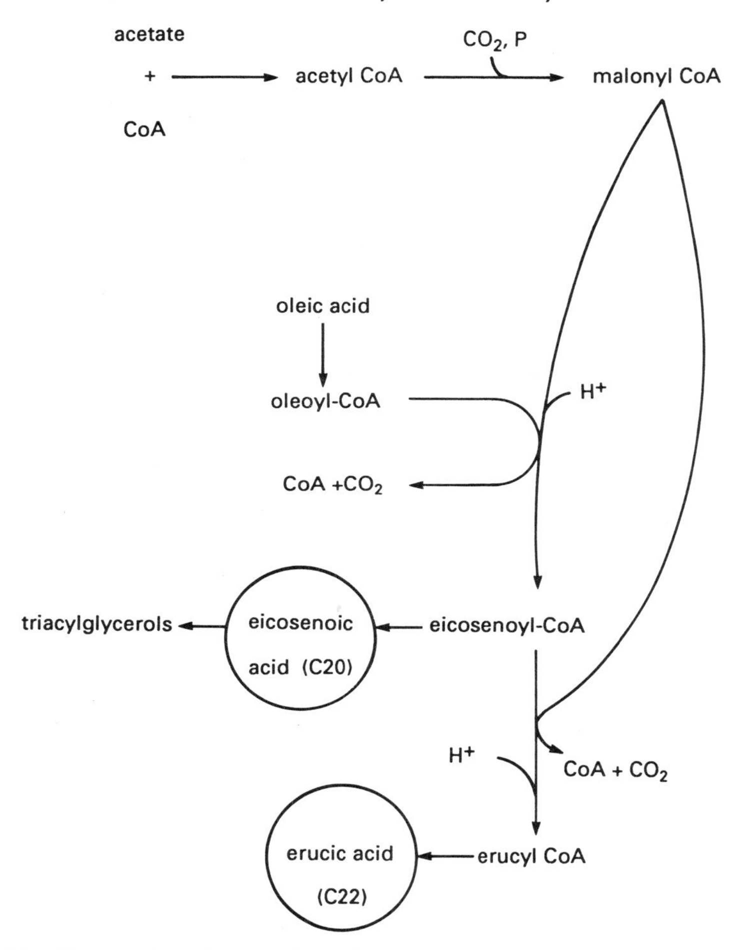

Fig. 7.10 Proposed pathway of synthesis of the long-chain fatty acids eicosenoic and erucic acids.

period of active triacylglycerol biosynthesis, in combination with enzyme purification studies. An alternative approach is to generate and characterize mutants with enhanced or decreased levels of stored lipids. The 'wrinkled' mutant of pea is known to reduce the starch content of the seed, but in some varieties at least also results in a slightly enhanced lipid content (although this may be simply compensatory). Nevertheless, this type of mutant may be a good target for genetic engineering studies. Interestingly, a recent study by Bayley *et al.* (1988) has shown that by transforming mouse fibroblasts with the gene for a rat medium chain hydrolase (an accessory enzyme of the fatty acid synthetase complex), there was produced an accompanying shift in fatty acid synthesis towards medium chain length. A similar experimental approach in plants would be valuable.

Technological Quality

The quality of crop products can be described not only in nutritional terms, but also in terms of technological and industrial requirements for the processing and marketing of those products. Two well-studied examples, in which a significant amount of information concerning the biochemistry and molecular biology of the systems is available, are bread-making with wheat flour, and the malting of barley and brewing of beer. Both processes are at least partly dependent upon the properties of the grain storage products, and manipulation of their chemical composition is expected to improve processing.

BREAD-MAKING

The bread-making quality of wheat depends primarily on the properties of the dough which allow it to rise and maintain its structure on baking, properties which are unique to wheat and are determined by the composition of the seed storage proteins. If starch, albumins, and other soluble components and globulins are washed out of the wheat dough, the resultant material, which possesses viscoelastic properties, is called gluten (Fig. 7.11). Gluten comprises two major protein (prolamin) fractions, gliadins and glutenin, constituting approximately 50% of the total grain protein. It is the gliadin fraction which is responsible for the viscosity of dough, allowing it to rise, while the glutenin fraction determines the elastic properties which prevent dough becoming over-extended and collapsing during the rising and baking processes, apparently as the result of interactions between high-molecular-weight (HMW, 95–150 kilodaltons) subunits of glutenins, via disulphide linkages. Both gliadins and the glutenins are the products of multigene families, and it is suggested that genetically determined differences in either the number and types of specific polypeptides (especially HMW subunits), in turn contributing to differences in chemical interactions between the polypeptides, account for the differences in baking quality observed with different wheat varieties. Therefore, the genetic manipulation of the proportion of specific HMW glutenins in wheat grain, and so of cross-linking polypeptides by altering cysteine residues at specific sites, has been proposed as a general strategy for improving baking quality. Before such objectives can be achieved, however, the precise relationship between amino acid sequence, cross-linking and the physicochemical properties of the proteins must be established.

BREWING

The production of beer and lager comprises two main processes: malting and brewing. For most European and North American beers, barley is malted, while African beers commonly use *Sorghum*. In the malting process, the cereal grain is germinated and then dried. It is mixed with other cereals, such as maize, rice or wheat (to form a 'grist'), before being mashed in warm water. During mashing, stored starch and protein are mobilized to sugars and flavour components by the action of barley amylolytic and proteolytic enzymes to produce a 'sweet wort'.

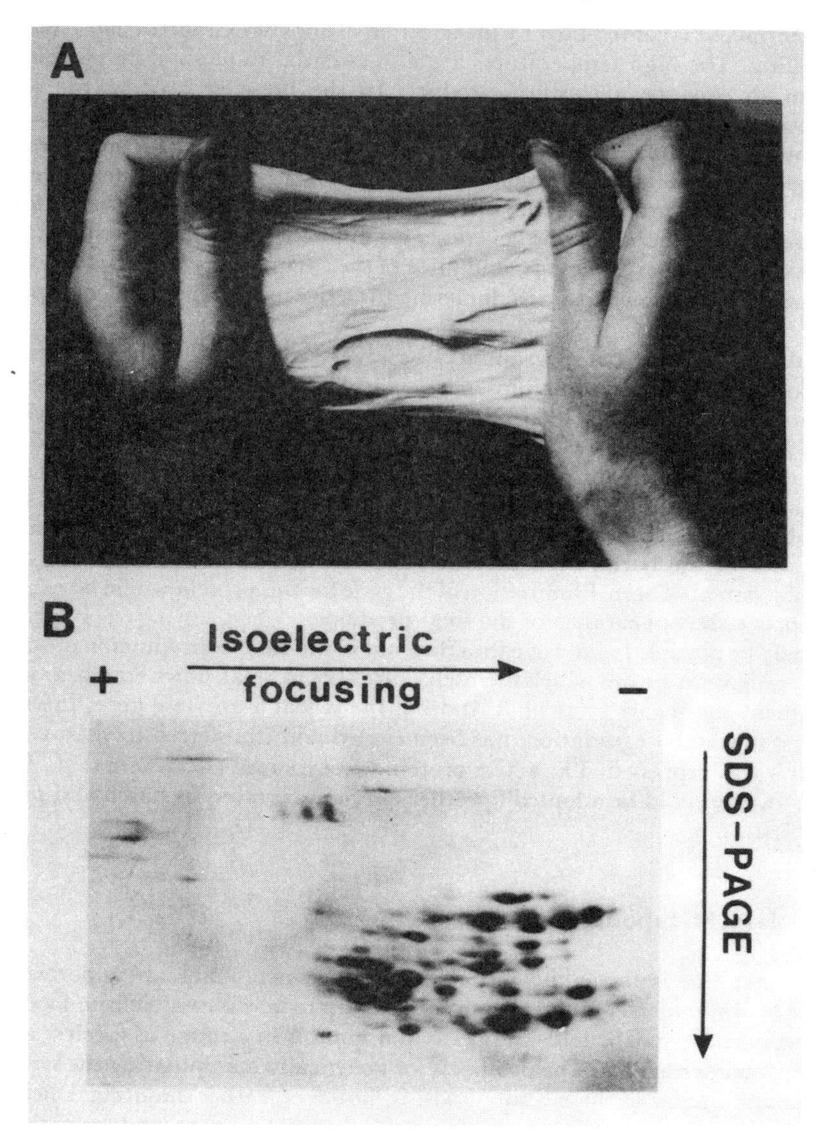

Fig. 7.11 Wheat gluten: (A) physical appearance and (B) component polypeptides separated by isoelectric focusing and electrophoresis (photograph kindly provided by Dr P.R. Shewry).

Further flavour is contributed by the addition of hops or extracts of hops, followed by boiling. The high temperatures are also essential to precipitate protein and tannins to produce a clarified product. In the brewing process, the sugars produced by malting are fermented to alcohol, and different species of yeast are employed according to the product. For example, ale production uses *Saccharomyces cerevisiae* strains, which are cultured on the surface of the fermenting liquor. For lager production *S. carlsbergensis* is used, which sinks to the bottom of the fermenter. Other alcoholic beverages use different substrates for fermentation, such as white grapes for cognac and juice of the cactus *Agave tequilana* for tequila, and specific yeasts, such as *S. saki*, for fermenting rice in the production of Japanese sake.

Malting quality of barley grain may be adversely affected by the amounts and composition of hordein prolamins, particularly B hordeins and disulphide-linked compounds (Miflin *et al.*, 1983). Such polypeptides appear to bind to starch and limit access by amylases. Some hordeins may also be degraded less during fermentation and create problems in other aspects of the brewing process in relation to filterability, foaming and cloudiness as a result of interactions with polyphenols. Beer haze is also produced by a protein, antigen I, which is a close derivative of the barley seed albumin, protein Z, and constitutes 10% of the protein content of beer. Elimination of the gene for this protein would be expected to improve the appearance of the final product.

It may be possible to improve the efficiency of the alcohol production process by developing yeast strains which themselves are able to break down starch, as well as to ferment the sugars to alcohol. Indeed, the wheat α-amylase gene, the major enzyme of starch degradation, has been cloned and transferred to yeast cells, in which it was expressed. The active protein was exported (Rothstein *et al.*, 1984). This strategy could be adopted for other enzymes involved in starch and protein hydrolysis.

Secondary Metabolites

In Chapter 4 we indicated how specific secondary metabolites are important as a source of economically useful chemicals as toxins to be eliminated from foodstuffs (e.g. steroidal glycoalkaloids in potato and tannins in a range of species) and as plant protective agents. The prospects for genetically manipulating the synthesis and accumulation of these compounds, is, however, rather daunting, since very little information is available concerning the regulation of secondary pathways, except in a small number of systems, and for many examples, the precise metabolic steps have not been completely determined. It is only possible, therefore, to suggest general strategies to increase or decrease the content of specific, or groups of related, compounds, strategies which may be effectively adopted only when key regulatory enzymes are identified and isolated and their genes have been cloned.

It has been argued that the expression of specific secondary metabolic pathways is a facet of differentiation, and is often linked to some aspect of structural development such as the morphogenesis of roots, leaves or fruits. For example, nicotine and related alkaloids are synthesized in the root system of the tobacco plant, although they may be transported upwards and accumulate predominantly in the aerial parts. Similarly, capsaicin, which gives the pungency of the chilli pepper *Capsicum frutescens*, is synthesized and accumulated at a particular stage of fruit development, and thte key enzyme appears to be active only in the placental tissue. The synthesis and accumulation of secondary compounds in cultured cells may largely be limited by a lack of differentiation, and the molecular biology of the regulation of capsaicin production has been speculated upon (Lindsey, 1988).

A second general feature of the regulation of the synthesis of at least some secondary metabolites is their inducibility, either in the intact plant or *in vitro*, and may be related to possible roles as protective agents. For example, the activities of enzymes involved in phenylpropanoid metabolism can be induced by environmental stimuli, such as blue or ultra-violet light, or by fungal elicitors. This phenomenon will be discussed in more detail in Chapter 9 in relation to the resistance of plants to disease and stress in the environment, but both the inducibility of specific enzymes and their genes, and their 'induction' in specific tissues or at particular developmental stages, provide a way to study key regulatory steps and to isolate the genes involved. It is known that plant tissues accumulate mRNAs which are specific to them, indicating differential transcription of the genome in different tissues, and approximately 0.1–1% of the total polysomal poly(A^+) mRNAs isolated were specific to a particular organ. Furthermore, *in situ* hydribization experiments demonstrated that some organ-specific transcripts may be restricted to certain cell types within an organ. As described in Chapter 3, developmentally regulated, as well as tissue-specific mRNAs, can be isolated and cDNA copies used to pick out specific genes from a genomic library. It should be possible, therefore, to isolate and sequence, and subsequently to modify the regulatory sequences of, genes for key secondary metabolic enzymes which are differentially expressed at the level of transcription. By changing sequences which determine, for example, tissue-specific expression for a constitutive promoter, such as that of the CaMV 35S RNA gene, and introducing the chimaeric gene into the plant, or simply into cultured cells, it may be possible to overcome the limitations of expression which exist in the wild-type plant and so produce a green bioreactor which accumulates enhanced yields of a desired metabolite. In support of this approach Nagy *et al.* (1985) have shown that the photoregulated expression of the small subunit of Rubisco can be overcome by replacing the natural promoter region with a sequence containing the CaMV 35S 5′ sequence in transgenic peas.

An alternative strategy might be to increase the activity of a transcriptionally regulated enzyme by introducing into the plant multiple copies of the gene. There is some evidence that, for example, the level of expression of the kanamycin-resistance phenotype in transgenic plants, encoded by a chimaeric neomycin phosphotransferase-II gene, is influenced by the copy number of the introduced sequence. However, it is generally recognized that position effects strongly affect the level of expression, and the targeting of genes to highly transcribed regions of the genome may prove to be a valuable, if difficult, approach.

It may also be possible to manipulate post-translational regulation of enzyme activity by modifying the interaction of enzyme and substrate. This could conceivably be achieved by increasing the affinity of the enzyme for substrate, but such a protein engineering strategy requires detailed knowledge of reaction mechanisms and a functional analysis of the enzyme polypeptide structure. It is also possible that the activity of a key enzyme is limited by the availability of substrate, either as the result of physical separation by compartmentalization, or because precursors to the desired product are utilized in competing metabolic pathways. It has been suggested, for example, that the synthesis and accumulation of capsaicin in cultured cells of the chilli pepper is limited, in part at least, by the effective reduction in substrate availability due to the 'sequestration' of aromatic precursors in protein and cell wall metabolism. It may be possible to overcome compartmentalization problems by targeting key enzymes into specific cell organelles, by linking specific transit peptides to the N-terminal region of the protein (see Chapter 3), but such an ambitious objective would require detailed knowledge of the subcellular localization of enzyme and substrate pools. Alternatively, it may be possible to reduce the activity of competing metabolic pathways to increase the supply of substrate to the terminal enzyme. An example, discussed in more detail by Lindsey (1988), is the synthesis of capsaicin. It would obviously not be feasible simply to eliminate the genes for the enzymes of the competing cell-wall biosynthetic pathways, but it might be possible to block the synthesis of one or more of the enzymes by the use of anti-sense RNA (aRNA, see Chapter 3). Anti-sense RNA binds specifically to mRNA ('sense' RNA), preventing its translation (see Fig. 3.13), and is a natural regulator of microbial gene expression. It has been further demonstrated that artificially synthesized aRNA is capable of inhibiting the synthesis of specific proteins in eukaryotic cells. In order that cell-wall synthesis would not be constitutively blocked in plants transformed with an anti-sense gene for the target enzyme, it would be desirable to link the chimaeric gene to an inducible promoter, such as that of a heat-shock gene.

A less sophisticated approach to increasing the availability of substrate is by the use of mutants. We have already seen how mutant cell lines, which over-produce amino acids, may be isolated from a population of cultured cells grown in the presence of a toxic analogue. Since many secondary products are derived from amino acids, such mutants might be expected to synthesize enhanced yields of a desired product, due to the increased flux of precursors. It has indeed been shown that tobacco cells, resistant to the toxic phenylalanine analogue *p*-fluorophenylalanine, accumulate increased levels of phenolics, concomitant with a 10–20 fold increase in the activity of the key enzyme phenylalanine ammonia lyase (Berlin and Widholm, 1977).

All the speculative strategies described in this section require that key regulatory enzymes can be identified and isolated. Therefore, the major hurdle in the genetic engineering of secondary metabolism, in which the desired compound is not a protein but the product of a complex pathway or set of pathways, is the requirement for much more basic biochemical information, rather than novel genetic manipulation techniques.

A technique which has much potential for the production of certain secondary

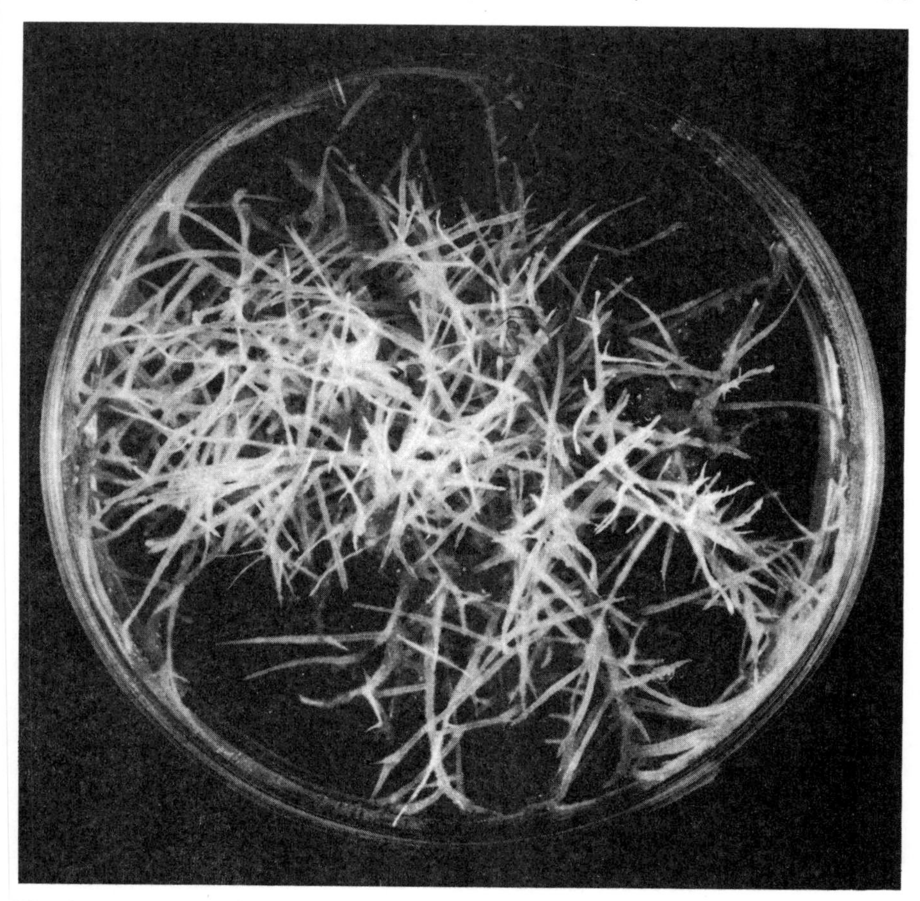

Fig. 7.12 Hairy root cultures of clover transformed by *Agrobacterium rhizogenes*.

metabolites *in vitro*, and exploits both the relationship between structural organization and product accumulation and also the characteristic properties of the *Agrobacterium rhizogenes* transformation system, is the use of hairy root cultures. As described in Chapter 5, wild-type *A. rhizogenes* carries a plasmid, the Ri plasmid, which induces the development, on inoculated cells, of numerous hairy roots (Fig. 7.12). Such roots can be cultured and, unlike undifferentiated culture cells, stably maintain the biosynthetic characteristics of the original plant. This system has therefore been used as a means of culturing cells which will synthesize and accumulate secondary metabolites characteristic of the roots of the intact plant. Members of the Solanaceae, which synthesize a range of valuable compounds in the roots, including tropane and nicotine-related alkaloids, have been studied in most detail to date (Flores *et al.*, 1987), but the technique will no doubt be applied to other plant families and secondary compounds.

General Reading

Hall, J.L., Flowers, T.J. and Roberts, R.M. (1981). *Plant Cell Structure and Metabolism*, 2nd ed. London, New York, Longman.
Steinback, K.E., Arntzen, C.J., Bogorad, L. (Eds) (1985). *Molecular Biology of the Photosynthetic Apparatus*. New York, Cold Spring Harbor Laboratory.
Stewart, W.D.P. and Gallon, J.R. (Eds) (1980). *Nitrogen Fixation*. London, Academic Press.

Specific Reading

Bayley, S.A., Moran, M.T., Hammond, E.W., James, C.M., Safford, R. and Hughes, S.G. (1988). 'Metabolic consequences of expression of the medium chain hydrolase gene of the rat in mouse MIH 3T3 cells', *Bio/Technol.* **6**, pp. 1219–1221.
Berlin, J. and Widholm, J.M. (1977). 'Correlation between phenylalanine ammonialyase activity and phenolic biosynthesis in a *p*-fluorophenylalanine-sensitive and -resistant tobacco and carrot tissue cultures', *Plant Physiol.* **59**, pp. 550–553.
Bright, S.W.J., Lea, P.J., Arruda, P., Hall, N.P., Kendall, A.C., Keys, A.J., Kueh, J.S.H., Parker, M., Rognes, J.E., Turner, J.C. and Wallsgrove, R.M. (1984). 'Manipulation of key pathways in photorespiration and amino acid metabolism by mutation and selection', in *The Genetic Manipulation of Plants and its Application to Agriculture*, Eds Lea, P.J. and Stewart, G.R., pp. 73–169. Oxford, Oxford University Press.
Bright, S.W.J. and Shewry, P.R. (1983). 'Improvement of protein quality in cereals', in *Critical Reviews in Plant Science*, Vol 1, Ed. Conger, B.V., pp. 49–63. Boca Raton, Florida, CRC Press.
Campos, F., Padilla, J., Vasquez, M., Ortega, J.L., Enriquez, C. and Sanchez, F. (1987). 'Expression of nodule-specific genes in *Phaseolus vulgaris* L.', *Plant Mol. Biol.* **9**, pp. 521–532.
Flores, H.E., Hoy, M.W. and Pickard, J.J. (1987). 'Secondary metabolites from root cultures', *Trends Biotechnol.* **5**, pp. 64–69.
Gonzales, R.A. and Widholm, J.M. (1985). Altered amino acid biosynthesis in amino acid analog- and herbicide-resistant cells', in *Primary and Secondary Metabolism of Plant Cell Cultures*, Eds Neumann, K.-H., Barz, W. and Reinhard, E., pp. 337–343. Berlin, Springer-Verlag.
Keys, A.J. (1983). 'Prospects for increasing photosynthesis by control of photorespiration', *Pestic. Sci.* **19**, pp. 313–316.
Keys, A.J. (1986). 'Rubisco: its role in photorespiration', *Phil. Trans. R. Soc. Lond.* **B313**, pp. 325–336.
Kuhlemeier, C., Green, P.J. and Chua, N.-H. (1987). 'Regulation of gene expression in higher plants', *Ann. Rev. Plant Physiol.* **38**, pp. 221–257.
Lindsey, K. (1988). 'Prospects for the genetic manipulation of complex metabolic pathways', in *Manipulating Secondary Metabolism in Culture*, Eds Robbins, R. and Rhodes, M.J.C., pp. 123–135. Cambridge, Cambridge University Press.
Miflin, B.J., Field., J.M. and Shewry, P.R. (1983). 'Cereal storage proteins and their effects on technological properties', in *Seed Proteins*, Eds Daussant, J., Mosse, J. and Vaughn, J., pp. 255–319. London, Academic Press.
Nagy, F., Morelli, G., Fraley, R.T., Rogers, S.G. and Chua, N.-H. (1985). 'Photoregulated expression of a pea *rbcS* gene in leaves of transgenic plants', *EMBO J.* **4**, pp. 3063–3068.

Prakesh, R.K. and Cummings, B. (1988). 'Creation of a novel nitrogen-fixing actinomycetes by protoplast fusion of *Frankia* with streptomyces', *Plant Mol. Biol.* **10**, pp. 281–289.

Rothstein, S.J., Lazarus, C.M., Smith, W.E., Baulcombe, D.C. and Gatenby, A.A. (1984). 'Secretion of wheat α-amylase expressed in yeast', *Nature* **308**, pp. 662–665.

Shewry, P.R., Williamson, M.S. and Kreis, M. (1987). 'Effects of mutant genes on the synthesis of storage components in developing barley endosperms', in *Developmental Mutants in Higher Plants SEB Semin. Ser.*, Vol. 32, Eds Thomas, H. and Grierson, D., pp. 95–118. Cambridge, Cambridge University Press.

Shinozaki, K., Yamada, C., Takahata, N. and Sugiura, M. (1983). 'Molecular cloning and sequence analysis of the cyanobacterial gene for the large subunit of ribulose-1,5-bisphosphate carboxylase/oxygenase', *Proc. Natl. Acad. Sci. USA* **80**, pp. 4050–4054.

Timko, M.P., Kausch, A.P., Hand, J.M., Cashmore, A.R., Herrera-Estrella, L., van den Broeck, G. and van Montagu, M. (1985). 'Structure and expression of nuclear genes encoding polypeptides of the photosynthetic apparatus', in *Molecular Biology of the Photosynthetic Apparatus*, Eds Steinbeck, K.E., Arntzen, C.J. and Bogorad, L., pp. 381–396. New York, Cold Spring Harbor Laboratory.

Twell, D. and Ooms, G. (1987). 'The 5′ flanking region of a patatin gene directs tuber-specific expression of a chimaeric gene in potato', *Plant Mol. Biol.* **9**, pp. 345–375.

Witty, J.F., Minchin, F.R., Skot, L. and Sheehy, J.E. (1986). 'Nitrogen fixation and oxygen in legume root nodules', *Oxf. Surv. Plant Mol. Cell Biol.* **3**, pp. 275–314.

Chapter 8

Manipulation of Reproductive Biology and Development

A thorough understanding of plant reproductive biology is essential for the development of successful breeding programmes and crop production. Some obvious aims of the manipulation of reproduction include the engineering of cytoplasmic male sterility, overcoming incompatibility barriers, higher yield and improved quality of seed, improved fruit set and quality, and so on. The principal features of seed development, in the broadest sense, are summarized in Fig. 8.1. The reproductive process from pollination to embryogenesis to seed germination and back to flower morphogenesis can be thought of as a cycle in which the distinct developmental stages are activated by the interaction of specific, but complex, physical and chemical stimuli with appropriately receptive tissues. The 'receptivity' of the tissues is as critical as the nature of the stimulus. For example, incompatibility mechanisms determine whether a pollen grain will 'germinate' on the surface of the stigma; and the phytochrome pigment system determines, in part at least, whether a seed will germinate or a seedling will elongate in response to long-wavelength light. Various stages of the reproductive cycle can be manipulated, not only by breeding practices, but also by the exogenous application of growth regulators. A dramatic example is the effect of spraying auxins on to dioecious plants, such as hemp (*Cannabis sativa*), in which the male and female reproductive structures are found on separate plants: spraying male plants causes the development of female flowers. On the other hand, the treatment of monoecious cucumber (*Cucumis sativus*) plants with gibberellins (GA_3) increases the ratio of male to female flowers which develop. These examples serve to illustrate that reproductive systems are susceptible to crude manipulations, but the techniques of molecular biology have now created the possibility of a detailed

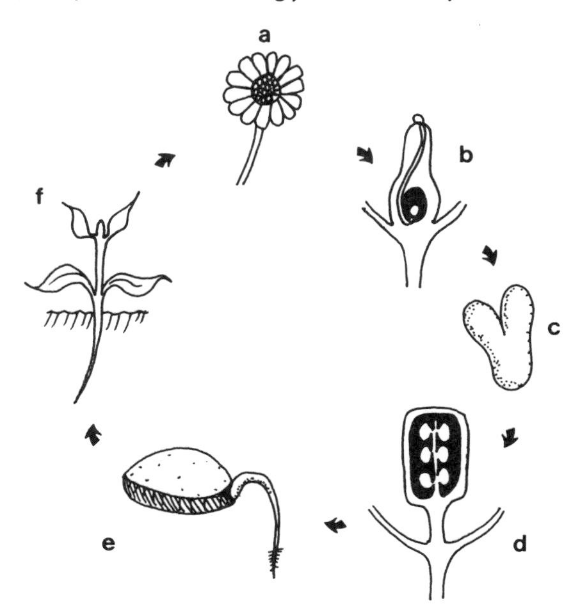

Fig. 8.1 Outline of plant reproductive development. (a) = flowering, (b) = pollination, (c) = embryogenesis, (d) = seed and fruit development, (e) = seed germination, (f) = seedling development.

understanding of the changes which occur during development, and of how they are regulated and may be altered, in very precise ways, to improve crop production and quality.

As indicated in Fig. 8.1, the main features of the reproductive cycle include: (1) the generation of gametes; (2) pollen–stigma interactions; (3) gamete–gamete interactions, leading to fertilization; (4) seed development; (5) fruit development; (6) seed germination; and (7) seedling development. The manipulation of fertilization is a fundamental concern of the plant breeder, and can be studied at three stages: the development of the pollen grains in the anther and the female gametes (the egg cell and central cell) in the ovule; the fate of the pollen grain once it has alighted on the surface of the stigma; and the transmission of the male nuclei to the egg and central cells.

Pollen Production

The haploid pollen grains are the products of meiotic cell divisions, and their subsequent development is characterized by the appearance of a number of specific transcripts (approximately 10–20% of the total pollen mRNA, according to the species), which do not appear in the diploid sporophytic tissue. The generation of cDNA clones coupled with differential hybridization has been used to pick out pollen-specific genes, which are thought to be involved in a whole range of functions, such as pollen development, germination, tube formation, sperm-cell formation, recognition of the stigma surface and penetration into the stylar tissues. Stinson *et al.* (1987), for example, have constructed cDNA libraries from poly(A^+)RNA isolated from mature pollen grains of maize and *Tradescantia paludosa*. Northern blot analysis indicated that there are at least two types of transcripts which differentially accumulate during pollen development. The first group, which were pollen specific, were synthesized after micropore mitosis, increasing in steady-state levels as the pollen matures. It is suggested that this group of mRNAs plays an important role in pollen grain germination and early growth. A second group of mRNAs, typified by the pattern of accumulation revealed by an actin cDNA probe, appeared soon after meiosis to reach a maximum level at late pollen interphase, followed by a decline in abundance, apparently due to non-specific degradation at the onset of pollen tube growth.

The study of the expression of specific pollen genes is in progress in the laboratory of Professor R.B. Knox (Melbourne, Australia); these are the *gal* gene controlling expression of β-galactosidase in pollen of *Brassica campestris*, and the *rye-1* gene which controls expression of the glycoprotein Group 1 Allergen in pollen of rye grass *Lolium perenne*. An understanding of the regulation of these and other pollen-specific genes would be expected to open up the possibility of manipulating the development and viability of pollen grains for traits such as male sterility and reduced allergenic properties.

The inability of some genotypes of certain species, notably *Sorghum*, maize, sunflower and sugar-beet, to produce viable pollen is a trait which is of interest to the breeder for the prevention of self-pollination, because it eliminates the labour-intensive necessity of removing the anthers by hand or by machine during the production of valuable hybrid seed. Since this trait is inherited in a maternal fashion from the female or seed parent, it has been termed 'cytoplasmic male sterility' (cms), and has aroused the interest of plant molecular biologists. Cms, in the species studied to date, is genetically determined by mitochondrial sequences, and the trait can be phenotypically reversed by so-called 'restorer' genes, which themselves are encoded in the nucleus. The mitochondrial genomes of higher plants are relatively large and can be represented as circles, ranging in size from approximately 200–2400 kilobases, according to the species. Plant mitochondrial DNA comprises about 50–55 genes, encoding the 26S, 18S and 5S ribosomal RNAs, at least 20 transfer RNAs and 18–20 polypeptides (see Leaver and Gray, 1982). Recombination between repeated sequences appears to give rise to smaller subgenomic circles, and there is now evidence that, in species such as maize, *Sorghum* and *Petunia*, specific sequence rearrangements are associated with the cms phenotype.

In maize, for example, cms is found in plants with particular cytoplasmic backgrounds, designated the C, T and S cytotypes. Normal male-fertile lines have the 'N' cytoplasm. The T ('Texas') cytoplasm is characterized by a sensitivity to the 'T-toxin' of the fungus *Helminthosporium* (*Bipolaris*) *maydis*, which causes

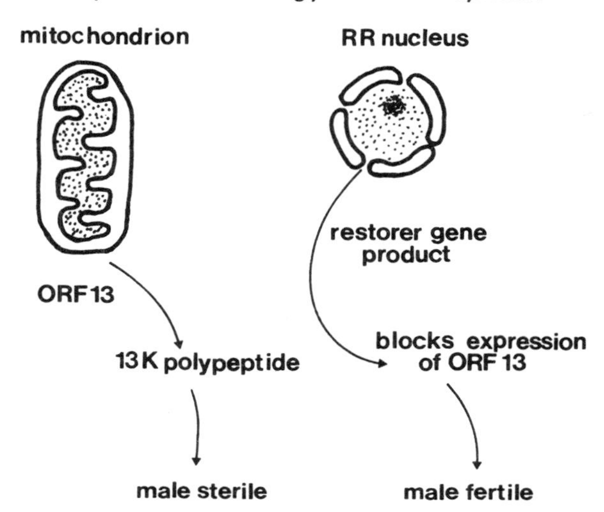

Fig. 8.2 Cytoplasmic male sterility in T cytoplasm maize. Male sterile lines are characterized by the presence of a mitochondrially-encoded 13-kDa polypeptide. Nuclear restorer genes (RR) interfere with the expression of the gene encoding this polypeptide, resulting in male fertility. See text for details.

southern corn leaf blight and which induces various abnormal symptoms in the mitochondria: swelling, leakage of NAD^+, an uncoupling of oxidative phosphorylation and ultimately an inhibition of respiration. T-cytoplasm mitochondria are also characterized by the presence of a novel 13-kilodalton polypeptide, located in the mitochondrial membrane fraction of the cell and which has now been shown to be synthesized by a novel open-reading frame designated ORF-13. ORF-13 is located within a chimaeric DNA sequence called TURF 2H3, which itself is the product of at least seven random recombinational events in the mitochondrial genome, and contains sequences which include part of the ATPase 6 gene, a region 3′ to 26S ribosomal DNA, and an Arg transfer RNA from the chloroplast genome. The ORF-13 gene and its product have been directly implicated in both the cms phenotype and also in mitochondrial sensitivity to the T-toxin. An antibody raised against a synthetic oligopeptide corresponding to part of the carboxy terminus of ORF-13 will immunoprecipitate the 13-kDa variant polypeptide which is synthesized by mitochondria from T-cytoplasm maize. Mutations in the ORF-13 coding region, or its deletion from the mitochondrial genome, are associated with a return to cytoplasmic fertility; transformation of *E. coli* with ORF-13 induces respiratory inhibition of the bacteria in the presence of the T-toxin (Dewey *et al.*, 1988). If, however, a mutated ORF-13 is used to transform, no toxin sensitivity is observed. Restorer genes suppress the level of the variant 13-kDa polypeptide, perhaps by altering the processing of the ORF-13 RNA (Fig. 8.2).

Mitochondrial DNA from C male-sterile cytoplasm also contains a novel recombinant open-reading frame, considered to be a chimaeric gene. Such mitochondrial genomes contain mutations in two ATPase genes (*atp9* and *atp6*) and in the cytochrome oxidase gene (*coxII*), which result from recombination events involving sequences of the mitochondrial genes and chloroplast DNA. It is believed that one of these mutations may be the cause of C-type cms, but this is yet to be confirmed. Mitochondria from S-cytoplasm maize contain two linear plasmids, S1 and S2 of 6.4 and 5.4 kilobases, respectively, which make up 5–15% of the mitochondrial DNA (mtDNA). These plasmids have terminal inverted repeats, and can integrate into the mitochondrial genome. When S1 and S2 are free, the mitochondria characteristically synthesize novel proteins, but the mechanism governing the return to male-fertility in such cytotypes is not precisely understood.

The influence of the nucleus on cms is exemplified by alterations to the structure of the nuclear-encoded gene for cytochrome oxidase subunit I (*CO I*). A survey of variation and expression of male-fertile and male-sterile nuclear–cytoplasmic combinations in *Sorghum*, for example, have revealed rearrangements of 5′ and 3′ regions of *CO I*. In the male sterile '9E' cytoplasm, a 3′ extension of the *CO I* gene results in the synthesis of an abnormally large (42 kDa) but related form of the 'wild type' protein (38 kDa), but the precise function of such rearrangements in relation to cms is not yet known.

From the available evidence, it seems that while specific mechanisms may vary between species or between genotypes (or cytotypes) within a species, cms is associated with the synthesis of novel polypeptides, which are products of the reorganization of mitochondrial DNA to form chimaeric sequences with new open-reading frames. Restorer genes may allow a return to fertility by interfering with some aspect of the synthesis or activity of the novel polypeptides. The function of these polypeptides is unclear, but they may compete with related native polypeptides at specific sites in the mitochondrion, such as the inner membrane, and so cause organelle dysfunction at microsporogenesis. Why all mitochondria in male-sterile plants are not similarly inhibited is a central question, but it may be that the mitochondrial abnormalities are only phenotypically revealed at developmental stages which require very high rates of mitochondrial activity or replication, such as at meiosis and pollen formation. The molecular study of cms is obviously of great potential for the modification of pollen production and aspects of disease resistance, and so of practical importance to the farmer in reducing the labour-intensive hand emasculation of male fertile lines through the genetic engineering of specific mitochondrial genes or of the nuclear restorer genes.

Pollen–Stigma Interactions

Once a pollen grain has alighted on the surface (pellicle) of the stigma, a series of molecular events is initiated, and if pollen and stigma are 'compatible', successful fertilization may occur. The term 'receptivity' is used to describe the capacity of

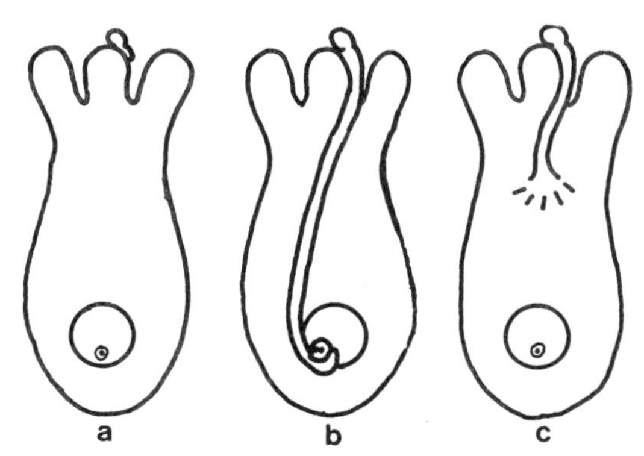

Fig. 8.3 Sporophytic and gametic systems of self-incompatibility. (a) sporophytic; (b) successful fertilization; (c) gametophytic.

the pistil to interact with the pollen, and is genetically determined. Receptive stigmata can recognize whether pollen of a different species will be accepted or rejected, and in approximately half of the angiosperm families self-incompatibility mechanisms have evolved to prevent self-fertilization. The genetics of self-incompatibility is well characterized, and two mechanisms, described as 'gametophytic' and 'sporophytic' systems, are illustrated in Fig. 8.3. In the gametophytic system, which is found in for example the Solanaceae, incompatible pollen germinates on the stigma, and the tube successfully penetrates the pellicle and grows some distance into the transmitting tissue of the pistil. However, the tip of the tube swells and bursts before reaching the female gamete, usually at a site which is characteristic of the species. For example, in the Gramineae, growth of the pollen tube is arrested within one or two cell layers from the point of entry through the pellicle, while in the cocoa plant (*Theobroma cacao*) the tube penetrates as far as the embryo sac but fails to enter the egg cell. In the sporophytic system of, for example, the Compositae and Cruciferae, incompatible pollen usually fails to germinate, or if it does, fails to penetrate the surface of the stigma. The genetic basis of self-incompatibility is at least partially understood. In the gametophytic system pollen rejection appears to be determined by the segregation to the male gametes of multiple alleles (designated S_1, S_2, etc.) of one or two major loci, S and Z. The important point is that each haploid pollen grain, carrying one of the alleles, *expresses its own phenotype* (i.e. whether it will germinate) determined by that single allele. Tobacco and rye exhibit this mechanism. In the sporophytic system, however, the phenotype (germination behaviour) is identical in all the pollen produced from a particular anther, irrespective of the segregation of the S alleles, and a dominant allele will determine the incompatibility response: in other words, it is the genes of the parent (sporophytic) tissue which define subsequent pollen activity rather than the genes of the individual (gametophytic) pollen grains themselves.

The precise role of the expression of these alleles in regulating pollen–stigma interactions is not clear. There are several stages at which molecular 'decisions' can be made once the microspore has arrived on the stigma. In a number of cases studied, the pollen-wall proteins have been found to bind to the proteinaceous pellicle. In compatible and in gametophytically incompatible interactions, hydration of the pollen and germination of the pollen tube occurs, while as we have seen, germination is inhibited in sporophytic or inter-specific incompatibility. It can be demonstrated by experiments in which non-viable compatible pollen or protein extracts of compatible pollen are applied to stigmata of normally self-incompatible species, that pollen-wall proteins, notably components of the sculptured exine, play a key role in the recognition reaction. In incompatible sporophytic matings, this may lead to the occlusion of stigma papillae and pollen tubes by callose.

It has been possible to identify at least some S-allele translation products as glycoproteins, by isoelectric focusing of stigma extracts. In sporotophytic *Brassica* species these glycoproteins have a molecular weight of approximately 55–65 kDa, and their role in self-incompatibility has been inferred through (1) the correlation of their inheritance with the S-allele segregation pattern, (2) their accumulation only in the stigmatal tissue, (3) their localization on the surface of the papillae of the stigma and (4) the correlation of their synthesis with the development of the incompatibility reaction. Stigma-specific mRNAs have been isolated and cDNA clones produced, which when cloned into an expression vector and translated in *E. coli*, produce a protein which was immunologically similar to a glycoprotein found in stigmata of plants bearing the S_6 allele. By nick-translating the cDNA to make a radioactive gene probe, it was possible to localize the transcripts, which were found exclusively in the surface layers of the stigmatal papillae (Nasrallah and Nasrallah, 1986).

In gametophytic self-incompatible species such as of the genus *Nicotiana*, glycoproteins associated with S locus alleles are localized along the length of the style, with the highest accumulated levels close to the point of pollen tube entry. cDNA clones of transcripts specific to particular glycoproteins have been made and used to localize S_2 mRNAs in the transmitting tissue of the style by *in situ* hybridization. The evidence presented illustrates the advances that are being made in our understanding of the regulation of pollen–stigma interactions, and the prospects for the isolation and manipulation of specific genes involved in the determination of tissue receptivity, a fundamental phenomenon in development, are now realistic. Perhaps the transfer of a self-incompatibility trait to self-pollinating species would be an alternative approach to engineering cytoplasmic male sterility to ensure the production of hybrid, rather than selfed, seed.

Gamete–Gamete Interactions

The fusion of the male and female gametes, fertilization, is the successful culmination of a series of recognition events leading to the setting of seed. The ovule (Fig. 8.4) comprises a number of cell types: the *egg cell*, surrounded by two *synergid cells* at the micropylar end, and a binucleate *central cell* (egg mother cell). Such species as *Plumbago* possess no synergids. At the chalazal end of the

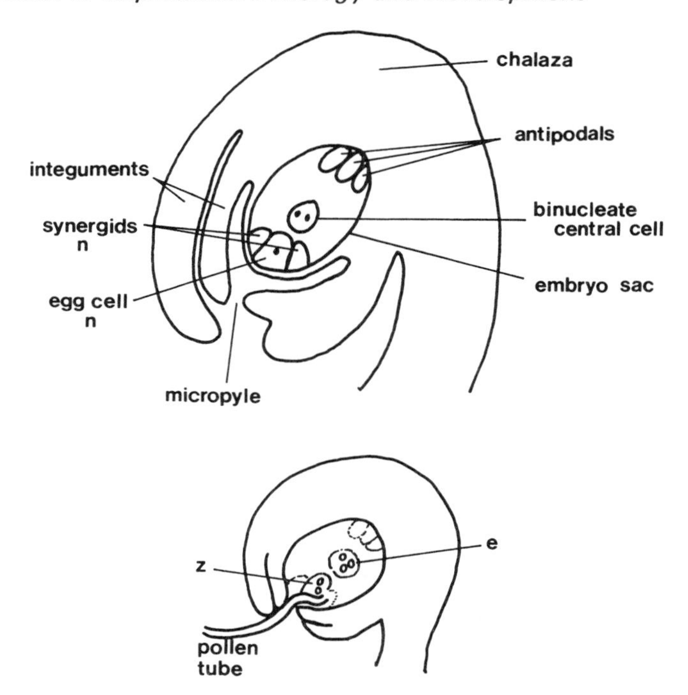

Fig. 8.4 (*Top*) Generalized structure of the ovule; (*bottom*) double fertilization occurs by the fusion of the sperm cells with the egg cell (to form the diploid zygotic nucleus, z) and the central cell (to form the triploid endosperm nucleus, e), respectively. The vegetative nucleus and pollen tube cytoplasm degenerate.

embryo sac are the *antipodal cells*. As the pollen tube reaches the ovule, the first microscopically observable effect is the structural breakdown of one of the synergids, and presumably a recognition system is involved in this process. On passing through the micropyle the pollen tube then pushes through the wall of the embryo sac and enters the degenerated synergid. Since it has been observed that the synergid is characterized by high levels of intracellular calcium, it is possible that the movement of the pollen tube is determined by a chemotrophic mechanism. The mechanism by which the two sperm cells and the vegetative nucleus move down the pollen tube is not known. Once in the synergid, the distal end of the pollen tube breaks down releasing the vegetative nucleus, the two sperm cells and a small quantity of cytoplasm. For a detailed review of the structure of the pollen tube cells, the reader is referred to Knox and Singh (1987). The subsequent processes constitute 'double fertilization' of the ovule, an important characteristic of the flowering plants (Fig. 8.4). The first event is the fusion of the nucleus of one of the sperm cells with the nucleus of the egg cell to produce the zygote which ultimately develops into the embryo. The second event is essentially a 'post-fertilization' fusion of the second sperm cell with the binucleate central cell to

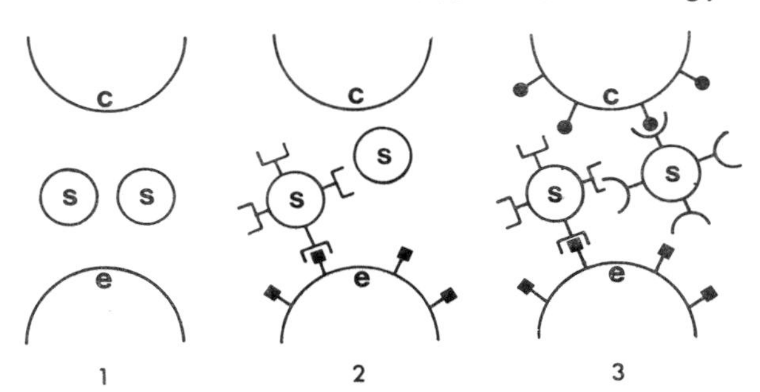

Fig. 8.5 Possible mechanisms of gamete–gamete recognition (after Knox and Singh, 1987). (1) The sperm cells (s) possess no specific recognition and fuse randomly with the central (c) and egg (e) cells. (2) Predetermined specific recognition of one sperm cell and the egg cell, leaving the second sperm cell to fuse with the central cell by 'default'. (3) Predetermined specific recognition between sperm cells and egg and central cells.

form a fusion product which develops into the triploid endosperm of the seed. The vegetative nucleus and pollen tube cytoplasm degenerate.

One major question arising from observations of double fertilization is whether the zygotic fertilization involves a specific sperm cell (a predetermined male gamete), or whether both sperm cells are potentially capable of fusing with the egg cell. There is currently little hard evidence which favours either possibility. It is possible that gamete–gamete interactions in plants are, in analogy to animal systems, determined by surface-recognition molecules, and Knox and Singh (1987) have discussed potential mechanisms (Fig. 8.5). One model proposes that the fusion of sperm and egg cells is a random event, but once fertilization has occurred, any further fusion events are blocked, preventing polyspermy perhaps by an alteration in the surface properties of the zygote. Such a model of random fusion requires no specific determinants on the sperm cell surface. If, however, fusion is non-random, there are two possibilities for the arrangement of recognition molecules: in the first, only the true male gamete is able to recognize and fuse with the egg cell, while the second sperm cell has no specific receptor molecules and therefore fuses with the central cell 'by default'. In the second possibility, each sperm cell possesses recognition molecules which are specific to the egg cell and central cell, respectively.

There is some evidence for preferential fertilization. For example, the sperm cells of maize exhibit genetic dimorphism, and it has been found that pollen from certain genotypes which carries non-disjunct B chromosomes is successful in fertilization at a relatively high frequency, suggesting preferential fertilization. It is not believed that the B chromosomes themselves determined the effect, since similar results were obtained in experiments using two different types of B chromosome. Support for a genetic recognition system has been provided by the observation that in *Plumbago*, which produces sperm cells rich in *either* mitochondria *or* plastids, there is preferential zygotic fertilization by plastid-rich sperm cells.

However, nothing is known about the possible recognition determinants in these or in any other experimental systems, although antibodies to sperm-specific surface proteins have been raised. *In vitro* systems will no doubt be developed to investigate the binding properties of isolated egg and sperm cells, and manipulation of these processes, and also of the genetic transformation of isolated gametes, e.g. by electroporation, is not a remote possibility. Although attempts to stably transform pollen have not yet met with success, it has been demonstrated that it is possible to increase the permeability of pollen to extracellular molecules by electroporation (Mishra *et al.*, 1987) and methods have been developed to isolate viable sperm cells (Dupuis *et al.*, 1987).

Seed Development

Fertilization of the ovule acts as a trigger for the further development of the seed, although in some species such as *Allium* spp., fertilization may not be an absolute requirement (a phenomenon known as 'agamospermy'). There are three essential features of seed formation: (1) development of the embryo; (2) development of the food reserves; and (3) development of the seed coat. In members of the Gramineae, for example, food reserves are accumulated in the endosperm ('endospermic seeds'), but in other species, such as sunflower and legumes, storage products accumulate in the cotyledons ('non-endospermic seeds'). Since little is known of the regulation of the development of the seed coat, this discussion of the biochemistry and molecular biology of seed development will be confined to embryogenesis and the synthesis of food reserves.

The development of the angiosperm embryo can be considered to occur in four parts: (1) the establishment of polarity, leading to the formation of root and shoot apices; (2) cellular differentiation; (3) desiccation and dormancy; and (4) germination. A number of biochemical, molecular biological and genetic techniques have been used to study embryogenesis both in the intact plant and *in vitro* with somatic embryogenesis, and can be categorized as studies of (a) the differential accumulation of a range of metabolites, notably proteins, lipids and carbohydrates, (b) the changing physiological activity, such as respiration and response to growth regulators, during development, (c) stage specific gene expression, and (d) developmental mutants. In relation to the possible manipulation of embryogenesis for plant breeding purposes, we will consider primarily the latter two approaches. For a review of the other features of embryogenesis, the reader is referred to Raghavan (1986).

The stages in the development of zygotic and somatic embryos of higher plants, exemplified by the carrot, are illustrated in Fig. 8.6. While the more advanced stages of development of somatic and zygotic embryos are comparable, there are certain differences which arise during the early divisions. After fertilization the zygote divides transversely, and the lower cell undergoes further divisions to produce the 'suspensor', distal to which is formed the embryo proper. The suspensor, the function of which appears to be to push the embryo up towards the developing endosperm, is usually not observed in somatic embryogenesis.

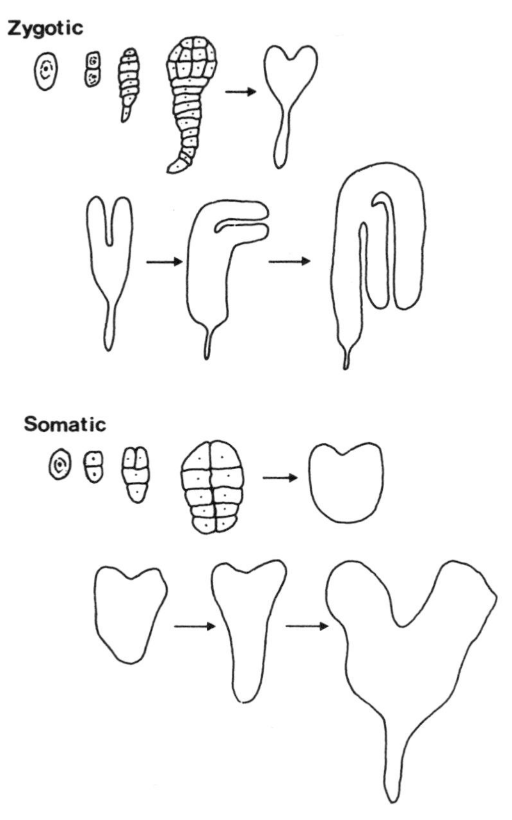

Fig. 8.6 Semi-diagrammatic representation of stages in zygotic and somatic embryogenesis of carrot. Individual cells are drawn for the earliest stages only.

However, the subsequent morphological stages of development are similar in both *in vitro* and *in vivo* systems, with characteristic globular, heart-shaped and torpedo-shaped phases. The pattern of further development of the cotyledons varies between species, according to whether the seed is endospermic or non-endospermic.

GENE EXPRESSION DURING EMBRYOGENESIS

The study of differential gene expression during embryogenesis requires the construction of gene libraries from specific developmental stages. This can be achieved by dissecting seeds at intervals after fertilization, or by isolating the different stages from *in vitro* cultures. Methods exist for enriching populations of cultured cells for specific stages, such as sieving techniques or the induction of some degree of synchronous development (Chapter 2). The objectives of molecular biological studies are commonly to identify and characterize genes which are expressed at specific stages, with the aim (a) of recognizing elements which control the establishment of polarity, the delineation of tissue layers and the segregation of meristematic apices and (b) of identifying markers which are characteristic of pro-embryonic cells (i.e. 'predictors' of embryogenesis). *In vitro* systems are heterogeneous in that a population contains both organized and unorganized tissues: the unorganized tissues can therefore be used as experimental controls, for they might be expected to express housekeeping, but not embryogenesis-specific, genes. A second approach would be to try and block putative key regulatory genes without inhibiting unorganized cell divisions, using anti-sense RNA.

The best evidence for the role of a gene in the regulation of a developmental pathway would be to reintroduce it into the same genotype after modification of its structure (e.g. mutate the coding region or alter the regulatory sequences), or into a mutant genotype which lacks it to produce a stably modified pattern of development of the transgenic embryo. To date, this has not been achieved, although stage specific mRNAs have been isolated and cDNAs produced (e.g. Dure, 1985). The growth regulatory substance abscisic acid (ABA) has been shown to influence the steady-state levels of particular stage-specific transcripts, which may be related to its effects on the precocious appearance of (late) development-specific polypeptides and on germination. However, it appears that ABA probably does not determine the *pattern* of embryogenesis, but may merely modulate the expression of certain genes and activity of gene products (Quatrano, 1986).

We can illustrate current experimental approaches by considering some specific examples of recent work. Using differential immunoadsorption techniques, Choi *et al.* (1987) have isolated rare, developmentally-regulated embryo-specific proteins from somatic embryos of carrot, and used them to isolate cDNA clones corresponding to the genes. By immunizing rabbits with a crude protein extract from unsynchronized embryos, an antiserum was obtained which recognized a large number of antigens from both structurally unorganized cultured cells and from somatic embryos. This suggests that the majority of the genes expressed in embryos were involved in cell division or 'housekeeping' functions rather than differentiation. However, rare antigens were detected which were specific to the embryo, and did not even persist in differentiated seedling tissues (leaves, petioles and roots). It was found also that the levels of the transcripts encoding the embryo-specific proteins were several-fold higher in the embryo than in the unorganized cells. The full length genes have not yet been isolated. Other tissues, namely somatic embryos and non-embryogenic callus of cassava, peach and maize, and zygotic embryos and leaf tissue of Johnson grass, were also tested for their reactivity with the embryo-specific antibodies. Although antibodies to two of the three carrot proteins failed to cross-react with extracts from any of the other tissues, it was found that all tissues produced a 50-kDa protein which was immunologically identical to one of the carrot antigens. Interestingly, the level of this protein was correlated with the ability of the other tissues to undergo embryogenesis, and is therefore worthy of further analysis.

The polarization of cell expansion and division is, as we have seen, a fundamental aspect of embryo development. It is believed that cortical microtubules may play a role in establishing cell polarity by determining the orientation of cellulose microfibrils in the cell wall. Indeed, the number of cortical microtubules increases dramatically during the progression of somatic embryogenesis in carrot. Cyr *et al.* (1987) have used a radioimmunoassay to demonstrate that, on a per-cell basis, the tubulin protein firstly decreases as undifferentiated cells begin to differentiate, but subsequently increases five-fold between the globular and torpedo/plantlet stages, concomitant with an increase in cell size. By using radioactively-labelled genomic sequences of the α- and β-tubulin genes of soybean as heterologous probes, it was demonstrated that steady-state mRNA levels for the carrot tubulin varied in accordance with the changing levels of the tubulin protein.

Very little work has yet been done to determine the fine control mechanisms of embryo-specific genes. However, efforts by Goldberg and co-workers have resulted in the identification of an embryo-specific DNA binding protein (see Chapter 3), which interacts with specific sequences in the 5′ region of a soybean lectin gene which is expressed during specific periods of embryogenesis and, in this case, also in the root of the mature plant. *In situ* hybridization experiments have shown that the lectin transcript is represented predominantly in the embryo cotyledon and in root ground meristem tissue. The DNA binding protein activity corresponds to a 60-kDa nuclear protein, and it is suggested that it, together with the sequences it binds to, are involved in the spatial and temporal expression of the lectin gene.

EMBRYO-LETHAL MUTANTS

An alternative approach to following stage-specific changes in gene expression is to characterize mutants which fail to develop normal embryos. Such mutants are described as 'embryo-lethal' and have been widely adopted in the study of animal development. In plants, the majority of the work in the genetic control of embryo development has been carried out with defective-kernel mutants of maize, embryo-lethal mutants of *Arabidopsis thaliana*, and variant cell lines of carrot which are unable to complete somatic embryogenesis (Meinke, 1986). Most of the defective-kernel mutants of maize have been obtained by the chemical mutagenesis of pollen, which is used to fertilize wild-type plants to produce recessive mutants which are detectable only after selfing: dominant lethal mutation-bearing individuals die before they can be analysed. Since the male and female structures of maize plants are derived from separate cell lineages in the embryo, mutagenesis of seed in this species is an unsatisfactory approach: a mutation in a single embryogenic cell will not be transmitted to both the male and the female gametes. This is not the case with *Arabidopsis*, where the entire shoot is derived from a single meristem, each cell of which can give rise to a number of individual flowers; and since pollen yield is small, seed mutagenesis is the method of choice. Variant cell lines of carrot, which are unable to develop mature somatic embryos, can be isolated without the use of mutagens by exploiting the natural heterogeneity that exists within a population of cultured cells. In particular, temperature-sensitive lines have been established which fail to develop fully when cultured at 32°C, although development is normal at 24°C. Since the precise genetic basis for this phenomenon is unclear, and there appear to be certain similarities to habituation (Chapter 2), the variant lines cannot strictly be described as mutants. However, their study is likely to provide information which will complement the work on lethal mutants.

A large number of embryo-lethal mutants have now been produced, and are described in some detail by Meinke (1986). The defective-kernel maize mutants fall into three categories: (1) those with an altered endosperm development, in which the embryos are nevertheless capable of maturing to form phenotypically normal plants; (2) those in which endosperm development is unaffected, but the embryos abort before they reach maturity; and (3) those in which the development of both endosperm and embryo is adversely affected, but is not necessarily lethal. However, although much information is now available concerning these mutants, little molecular biological analysis of them has so far been carried out. One hurdle to be overcome is the development of techniques for the genetic transformation of maize, to aid investigation of the functional significance of mutant genes, as described earlier. However, advances have now been made in methods for the regeneration of plants from maize protoplasts, opening up the opportunities for direct gene transfer.

Arabidopsis thaliana is probably much more amenable to molecular investigation, and has proved to be an excellent experimental system for a number of reasons. The life cycle of the plant is very short, of the order of 5–6 weeks, and the plant can be grown at high density on a defined nutrient medium, allowing the ready screening of large numbers of potential mutants. It is self-fertile, out-crossing at a low frequency (approximately 10^{-4}), so genetically stable lines can be maintained without difficulty. The genome is the smallest of the higher plants, with the haploid five chromosome complement comprising about 70 000 kilobase pairs, which is only five times larger than the genome of yeast. This is of particular advantage for gene isolation studies: only 16 000 λ clones, with average inserts of 20 kilobase pairs, need be screened in order to obtain a 99% probability of retrieving a specific sequence from a genomic library (Estrelle and Somerville, 1986). This compares with 310 000 clones necessary to achieve the same result with a genomic library of sugar-beet. The small size of the *Arabidopsis* genome appears to be due primarily to the small amount of repetitive DNA. A linkage map and a series of well-characterized mutants are available (Table 8.1), and tissue-culture systems have been developed.

Embryo-lethal mutants abort over the full range of stages from the zygotic to the cotyledonary stages of embryogenesis, and even at germination and seedling growth. Unusual mutant phenotypes have also been obtained, with excessively large suspensors, distorted and fused cotyledons, reduced hypocotyls, and arrested embryos without distinct cotyledons or hypocotyl tissue. Molecular analysis of such mutants may provide answers to a number of questions concerning the regulation of embryogenesis:

- how is suspensor development controlled?;
- how do the embryo and endosperm interact?;

Table 8.1 Some genetic markers of *Arabidopsis thaliana*

Name/symbol	*Phenotype*
Albina, *alb-1*	White embryos and seedlings (lethal)
Apetala, *ap-1*	No, or only rudimentary, petals
Apetala, *ap-2*	Reduced petals, large sepals
Chlorate resistance, *chl-1*	Chlorate resistance due to reduced uptake of chlorate
Chlorate resistance, *chl-2*	Chlorate resistance due to decreased nitrate reductase activity
Gibberellin-requiring, *ga-1*	Gibberellin-responsive dwarfs
Glabra, *gl-1*	Trichomes absent on leaves and stems
Pistillata, *pi*	Anthers and petals absent
Sulfurata, *su*	Bright yellow–green plant
Late flowering, f_b	Flowers later than wild-type, more rosette leaves
Thiamine-requiring, *th-i*	Leaves, except cotyledons, white–yellow, lethal, wild-type phenotype restored by thiamine

- how autonomous is the development of the embryo?;
- how many genes are critical for the development of the embryo?;
- how many genes are critical for the progress of specific phases of embryogenesis?;
- is zygotic and somatic embryogenesis controlled by the same genes?;
- can genes be isolated which may help overcome the barriers to successful embryo development in interspecific hybrids?

The use of transposon mutagenesis should prove to be a valuable technique, both for generating developmental mutants and for isolating the mutated gene, by probing extracted DNA with the transposon. Such an approach has already been demonstrated in animal systems, and awaits exploitation in plants.

STORAGE PRODUCTS

The molecular biology of seed storage product accumulation has already been discussed briefly in relation to improving nutritional and technological quality. Because of the complexity of their synthesis, via a number of metabolic steps, the regulation of gene expression in the developmental accumulation of lipids and carbohydrates has not been elucidated. Rather, most effort has been directed to the study of the synthesis of the seed storage proteins, and has provided a great deal of information on the biochemical differentiation of the endosperm and the cotyledons. Some storage proteins, such as the chymotrypsin inhibitors and α-amylase of barley, are synthesized in the embryo as well as the endosperm, but the barley hordeins, wheat α-gliadins and maize zeins are endosperm-specific. Hordeins, gliadins and zeins are classified as prolamins, i.e. single polypeptides which are soluble in alcohol, are rich in proline and glutamine and poor in basic amino acids such as lysine (see Chapter 7), and are found solely in the seeds of the Gramineae. Genetic and molecular biological studies have demonstrated that these and other storage proteins, such as the globulins of cotton (*Gossypium hirsutum*) and the glycinins and conglycinins of soybean (*Glycine max*), are encoded by multigene families. The wheat gliadin genes, for example, are located on different chromosomes and direct the synthesis of approximately 75 superabundant mRNAs (21 000 copies per cell), representing almost 60% of all transcripts which are translated. The gliadins are synthesized on membrane-bound polysomes as apoproteins carrying peptide signals, and are co-translationally modified as they are transferred to the lumen of the endoplasmic reticulum. The prolamins of barley comprise the B, C and D hordeins, representing the sulphur-rich, sulphur-poor and high-molecular-weight (HMW) proteins, respectively, and are encoded by three loci on chromosome 5. As indicated in Chapter 7, the levels of accumulated hordeins and their mRNAs are differentially influenced by various 'high-lysine' mutations, occurring at different loci to the genes of the proteins affected; for example, the *lys* gene of Hiproly is located on chromosome 7.

The co-ordinated spatial and temporal expression of the cereal prolamin genes during endosperm development is regulated in part, at least, at the level of transcription. Analysis of gene structure has revealed highly conserved sequences 5′ to the coding region and within 600 base pairs of the translation initiation codon. In particular, there has been found a short sequence, approximately 300 base pairs upstream of the ATG, which appears to be unique to the prolamin genes, and its functional significance has been analysed by the use of transgenic experimental systems. Colot *et al.* (1987) have linked 5′ sequences of the genes for two endosperm-specific storage proteins of wheat, a HMW and a low-molecular-weight (LMW) glutenin, respectively, to the chloramphenicol acetyltransferase (CAT) coding sequence, and introduced the chimaeric constructs into tobacco. Tobacco was used because transformation systems for cereals have not been fully developed. Both HMW and LMW constructs directed endosperm-specific CAT activity, and a series of deletion mutations in the upstream region of the LMW glutenin genes indicated that sequences between 326 base pairs and 160 base pairs 5′ to the transcription initiation site were necessary for tissue-specific expression. Similarly, Marris *et al.* (1988) have linked 549 base pairs of the 5′ flanking region of the barley B1 hordein gene, containing the so-called '−300 element', to the coding region of the CAT reporter gene, and introduced the chimaeric construction into tobacco plants using *Agrobacterium tumefaciens* as the vector. Analysis of the transgenic plants revealed that CAT activity was detectable only in the endosperm of the seed, and only from 15 days after pollination (Fig. 8.7). However, in plants transformed with a chimaeric gene comprising the 'constitutive' cauliflower mosaic virus 19S RNA gene promoter linked to CAT, activity was found throughout seed development in both embryo as well as in endosperm and also in leaf tissue. Chen *et al.* (1988) have used similar techniques to identify regulatory sequences in the gene for the α-subunit of β-conglycinin, an embryo-specific storage protein of soybean. An element between −143 and −257 base pairs determines tissue- and stage-specific expression of the gene in transgenic *Petunia* plants, and has further been shown to have an effect as an enhancer element, but in a tissue- and development-specific fashion when linked to a constituitively-expressed reporter gene (a CaMV35S promoter/CAT construction).

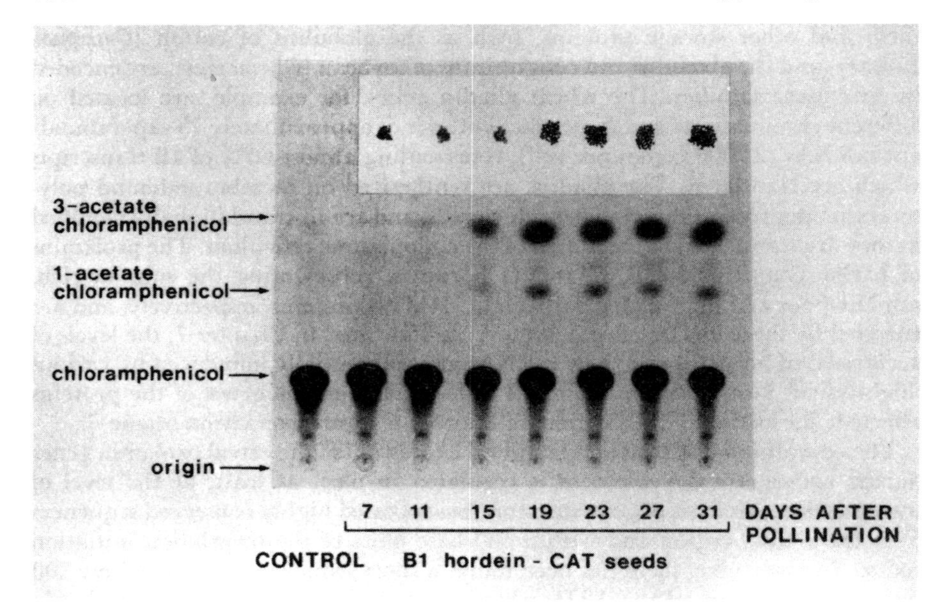

Fig. 8.7 Expression of a B-hordein promoter/CAT coding region gene fusion in developing tobacco seeds. CAT enzyme activity is under the control of the B-hordein promoter, which directs accumulation of the active protein only beyond 15 days after pollination. The same developmental control is exerted over the accumulation of the B-hordein storage protein. (Photograph kindly provided by C. Marris.)

These results therefore demonstrate how the orchestrated expression of genes, tightly linked to specific phases of seed development, is determined in part by *cis*-acting, non-transcribed, sequences. Progress is also being made in our understanding of how these sequences exert their importance through interactions with *trans*-acting DNA binding proteins; however, our knowledge of such interactions is still far from complete. Furthermore, factors other than those of transcriptional regulation appear to play a role in determining whether a protein, as opposed to a mRNA, will accumulate. Tercé-Laforgue *et al.* (1987) have found that, for the synthesis of gliadins in the wheat grain, there was little correlation between the steady-state levels of specific transcripts and their translation products, emphasizing the role of translational control of protein accumulation. Nevertheless, the prospects for manipulating aspects of the biochemical differentiation of seeds, if not yet of embryogenesis itself, are evident.

Fruit Development

During the development of the seed the ovary wall undergoes a series of changes culminating in the maturation of the fruit, which acts both to protect and to

methionine ⟶ S-adenosylmethionine $\xrightarrow[\text{ACC synthase}]{}$ ACC $\xrightarrow[\text{'EFE'}]{}$ ethylene

Fig. 8.8 A summary of ethylene biosynthesis. ACC = 1-aminocyclopropane-1-carboxylic acid; EFE = ethylene-forming enzyme.

propagate the seed. The precise pattern of fruit development varies widely between species, but the final product can broadly be categorized in morphological terms as a fleshy fruit or a dry-walled fruit. Most of the recent developmental biological studies have used fleshy fruits as experimental systems for a number of reasons: developmental mutants are available for some species; some fruits exhibit dramatic changes in metabolic physiology during their development; and many are of commercial value. Furthermore, much interest has been directed to the biochemical and molecular biological events of fruit ripening. For many years it was considered that the ripening process was essentially catabolic in nature, representing a gradual decay of the organized state. It is certainly true that there are aspects of senescence involved, but molecular biological techniques have revealed that ripening is under genetic control, and is characterized by the stage-specific synthesis of a range of molecules and the coordinated expression of specific genes (Brady, 1987). During the early phases of fruit formation there is a rapid increase in the rates of cell division and/or cell expansion, with the accumulation of storage materials such as starch or fats. The decline in the rate of growth signals the onset of maturation, and then ripening, which may be characterized by a change in pigmentation, the synthesis of specific secondary metabolites such as flavour or aromatic compounds or the breakdown of others, such as alkaloids, the degradation of starch to sugars, and softening of the tissues. Certain fruits exhibit an increase in respiratory activity during ripening, called the 'climacteric'. Species which show this effect, such as apples, tomatoes and bananas, have correspondingly been described as 'climacteric fruits', while other species such as oranges, pineapples and cherries are 'non-climacteric fruits'. Climacteric fruits also characteristically produce ethylene during ripening, and it now seems likely that it is ethylene which induces the increase in respiratory activity in these fruits. The exogenous application of ethylene can advance both the onset of ripening and also the induction of further ethylene biosynthesis, and control of the effects of ethylene is of great commercial interest, from the points of view both of hastening ripening prior to harvest and of the long-term storage of fruits.

A summary of the pathway of ethylene biosynthesis is illustrated in Fig. 8.8. Ethylene production is normally limited by the availability of its immediate precursor 1-aminocyclopropane-1-carboxylic acid (ACC), itself synthesized from S-adenosylmethionine via the regulatory enzyme ACC synthase. The activity of this enzyme rises during ripening, relieving the substrate limitation. However, when immature tomatoes are exposed to exogenous ethylene they fail to ripen, indicating that the 'receptivity' of the fruit to physiological signals varies through development. The molecular and biochemical bases of these changes are unclear, and may relate to interactions with other endogenous growth regulators or depend upon the pattern of expression of specific genes which has occurred at the time of ethylene application (Brady, 1987).

Table 8.2 Some ripening mutants of tomato

Name/symbol	*Phenotype*
Ripening Inhibitor, *rin*	Fruits exhibit no respiratory climacteric or ethylene production, and soften only slightly. Mature colour is yellow, and ripening is not restored by exogenous ethylene
Non-ripening, *nor*	Similar to *rin*, but mature fruit is orange
Neverripe, *nr*	Fruits soften slowly, and mature colour is orange
Alcobaca, *alc*	Fruits soften slowly, and mature colour is pale red
Yellow flesh, *r*	Fruit development is normal, except for the failure to accumulate lycopene. Mature colour is yellow
Green flesh, *gf*	Fruit development is normal, except for the incomplete loss of chlorophyll. Mature colour is red–brown
Tangerine, *t*	Fruit development is normal, except for the accumulation of prolycopene instead of lycopene. Mature colour is orange.

The genetic basis of ripening, and also of sensitivity to signal molecules, is demonstrated by a range of ripening mutants of the tomato (Table 8.2). Of particular interest in relation to the role of ethylene in ripening are the 'ripening inhibitor' (*rin*) and 'nonripening' (*nor*) mutants. Plants which are homozygous for either of these recessive mutations produce fruits which change colour (becoming yellow (*rin*) or pale orange (*nor*)), but exhibit no significant respiratory climacteric, ethylene production or wall-softening characteristic of the wild-type fruit. In the presence of exogenous ethylene, the respiration rates of both mutants increase, but the capacity to respond in terms of the other facets of ripening is absent. In *rin* mutants, exogenous ethylene induces the activity of the ethylene-forming enzyme (Fig. 8.8), but not of ACC synthase. Other mutants, such as 'neverripe', 'green ripe' and 'alcobaca', ripen very slowly or incompletely, and again exogenous ethylene fails to speed up the process. There is some evidence that water stress can partially induce ripening in *nor* mutants, suggesting that the metabolic machinery for ripening is not completely lost, but can be switched on under certain environmental conditions.

Although the molecular basis for the inability to respond to ethylene is unknown, other aspects of the molecular biology of tomato ripening have been characterized, although these appear to be largely symptomatic rather than 'causative' of the response. There is now a large amount of evidence from radioactive-labelling experiments that ripening fruits are capable of the synthesis of nucleic acids and proteins, and more recent studies demonstrate that this is partly a reflection of the differential expression of specific genes. In tomato, changes in the steady-state levels of mRNAs have been analysed by *in vitro* translation techniques, and three classes of transcripts have been identified. One class is present throughout fruit development and ripening, and is presumed to represent the products of housekeeping genes. A second class of mRNAs appears during the early stages of fruit development, but declines in abundance (or in translatability) before or at the onset of ripening. The third class, representing at least eight transcripts, appears predominantly during the ripening process. In this and in the other well-studied system, the avocado, relatively few of the transcripts are strictly ripening-specific, and the proteins they encode are largely unidentified. It is known that, during ripening, there are increases in the activities of the enzymes cellulase (in avocados), invertase and polygalacturonase (in tomatoes), but the identification of specific transcripts has only been achieved for avocado cellulase (Christoffersen *et al.*, 1984) and tomato polygalacturonase (Slater *et al.*, 1985) by immunoprecipitation of translation products of hybrid-selected mRNAs. Northern blotting demonstrated that the mRNA corresponding to the avocado cellulase cDNA clone was of approximately 2000 base pairs, and while it is barely detectable in the unripe fruit, its accumulation can be enhanced 50-fold by ethylene-induced ripening. The primary translation product is a 54-kDa preprotein from which a 1.2-kDa signal peptide is removed during membrane transport. In the membrane the protein is glycosylated and finally modified to produce a mature protein of 54.2 kDa associated with the cell wall where it is presumably involved in the softening process.

Tomato polygalacturonase is synthesized *de novo* during ripening *after* ethylene synthesis is initiated. Both the mature enzyme and the polygalacturonase mRNA are absent in green tomato fruits, but their appearance can, however, be induced by exogenous ethylene. In 'neverripe' mutants, which soften very slowly, relatively low levels of the enzyme accumulate compared with wild type fruits, while *rin* mutants produce none at all, and the polygalacturonase mRNA, which is also absent, is not induced in *rin* mutants by the application of ethylene.

Apart from the appearance of new transcripts during ripening, others (i.e. those of Class 2) disappear, and some have been identified. Included are transcripts of the chloroplast genome which code for photosynthetic enzymes, such as one for a protein of photosystem-I (*psaA*), four for photosystem-II (*psbA*, *psbB*, *psbC* and *psbD*), and one encoding the large subunit of Rubisco. Interestingly, the transcription of some of these and other plastid genes may be partially controlled by diurnal rhythms.

Although the control mechanisms of fruit ripening are unclear, and so genetic manipulation of development is some way off, it is apparent that molecular techniques are greatly aiding progress. The production of molecular probes to the genes of ACC synthase and the ethylene-forming enzyme will help in the elucidation of the regulation of ethylene biosynthesis, and sequence analysis of ethylene genes, followed by linking putative regulatory sequences to marker genes, should yield valuable information.

Seed Germination and the Mobilization of Food Reserves

After dispersal of the seed, germination will occur if certain criteria are met. If water is available the desiccated seed will imbibe, and in the presence of an adequate oxygen supply (to permit aerobic respiration) and high enough temperatures (to permit adequate rates of enzyme activity), germination will take place with the emergence of the radicle and the cotyledon(s). Seeds of species which live in a wet environment, such as rice (*Oryza sativa*) and bullrush (*Typha latifolia*), do not require high oxygen levels for germination. The early heterotrophic growth of the emerging seedling is dependent upon stored food reserves in the endosperm or cotyledons, which are mobilized by the action of enzymes; the food reserves are not thought to be essential for the germination process itself since embryos isolated from the endosperm may still germinate. It may be the case that germination of a viable (living) seed fails even in the presence of water, oxygen and warmth, and in this situation the seed is described as 'dormant'. Dormancy may, however, be overcome by specific environmental factors such as light or low temperatures, and once the plumule is pushed out of the soil autotrophic growth can commence.

Information on the molecular biology of germination itself is scarce, and therefore the identification of regulatory genes and the manipulation of the process is some way off. It is known that phytochrome is one factor to play a role in inducing the germination of light-sensitive seeds, and the role of this pigment in development is considered briefly later. The molecular basis of the effects of the vernalization of seed to promote germination is not well understood, but the salient features of the biochemical events of dormancy and the breaking of it are worthy of mention.

The completion of seed maturation is characterized by the rapid dehydration of the tissues, resulting in profound changes in subcellular organization, accompanied by a sharp decline in the rates of protein synthesis and respiration. It should be noted that water loss is not absolute, cereal grains having a 10–15% w/w moisture content even in the 'dry' condition. The mitochondria and other structures, such as cell nuclei, pastids, endoplasmic reticulum and Golgi apparatus, become indistinct as seen under the electron microscope, and there is a decrease in the activity of a range of specific enzymes, such as those involved in glycolysis, the pentose phosphate pathway, the tricarboxylic (TCA) cycle and the cytochrome system. Although in some seeds there may be a loss of activity of amino acid and protein synthetic machinery (in the dehydrated castor bean there is a breakdown of polysomes and ribosomes), this is not universal, since the mature wheat seed requires only translatable mRNA to carry out protein synthesis, and indeed has been widely exploited as an *in vitro* translation system (see below). A number of mRNAs (such as those of actin, tubulin and calmodulin) are stored throughout the period of dehydration, and some may be translated on imbibition (Dure, 1985). Studies with inhibitors of transcription have suggested that the steady-state levels of persistent transcripts are due to the stability of the mRNA rather than to a high rate of synthesis relative to degradation, and the precocious translation of the messages may be inhibited *in vivo* by abscisic acid (reviewed by Bewley and Black, 1978).

The imbibition of water is an essential first step in the return to an active seed metabolism and ultimately to germination. Hydration is usually accompanied by the re-activation of the respiratory pathways of glycolysis, the TCA cycle and the pentose phosphate pathway, and the deamination and transamination of stored amino acids to keto acids provides a source of substrate (e.g. as α-ketoglutaric acid or pyruvic acid). A rapid rise in the rate of oxygen consumption occurs in two phases, pre- and post-germination. Both ethanol and lactic acid may accumulate if oxygen availability is limited, and alchohol dehydrogenase, present in dry pea seeds, catabolizes the alcohol and then declines in abundance during aerobic development. The regulatory sequences of an alcohol dehydrogenase gene from maize have been characterized, and the inducibility of transcription demonstrated in conditions of low oxygen tension (see Chapter 9). Substrate, other than from amino acids, for early respiration is available predominantly as sugars, such as sucrose, raffinose (galactose : glucose : fructose) and stachyose (galactose : galactose : glucose : fructose) stored in the dry seeds. The major food reserves such as lipids and carbohydrates are utilized only later. In a number of species the catabolism of raffinose and stachyose is coincident with an increase in the activity of α- galactosidase, but the level at which this is regulated has not been determined. The appearance, under the electron microscope, of mitochondria changes dramatically during imbibition, with increases in the number of cristae per organelle, indicative of reorganized metabolic activity. In imbibed peanut cotyledons, mitochondrial DNA and protein synthesis occur, with concomitant increases in the activities of cytochrome *b*, cytochrome *c*, cytochrome oxidase, succinic dehydrogenase, succinoxidase and in, for example, the pea cotyledon, malate dehydrogenase. There appear to be two general patterns of mitochondrial development in imbibed seeds: the repair and re-activation of mitochondria present in the dry seed (the situation found in the pea cotyledon), and the biogenesis of completely new organelles (as seen in the peanut cotyledon). This aspect of development is considered in more detail by Bewley and Black (1978, 1985).

Protein synthesis is also rapidly induced on imbibition. Polysomes, absent in dry seeds, appear within about 15 min. of water uptake in wheat embryos. As indicated earlier, cytoplasmic fractions of dry wheat embryos, or dry peanut seeds, can support the *in vitro* translation of polysomes, and therefore must contain the essential protein synthetic apparatus, including tRNAs, amino acids, translation initiation and elongation factors, and the supporting enzymes. Synthesis of ribosomal and transfer RNAs (rRNA, tRNA), however, is also detectable within a short period after imbibition.

Once germination is initiated, the early, heterotrophic growth of the seedling is dependent upon an energy supply from the food reserves stored in the cotyledons or the endosperm. Seeds can be classified according to the nature of the energy-rich reserves, i.e. fat-storing or starch-storing. The mobilization of starch during the germination and early development of the barley seedling is a well-studied system. Interest arises from the commercial value of the malting process already discussed (Chapter 7), and from the scientific viewpoint, because the control mechanism represents a now classic example of the interaction of a plant growth regulator with a specific tissue type to elicit the activity of a specific gene or genes. Whether this represents the typical mode of action of a growth regulator as a 'hormone' is a separate question altogether, and we will not attempt to answer it (but see Guern, 1987).

The structure of the barley seed is illustrated in Fig. 8.9. The aleurone layer is characterized by the thick-walled cells containing granules of protein (the

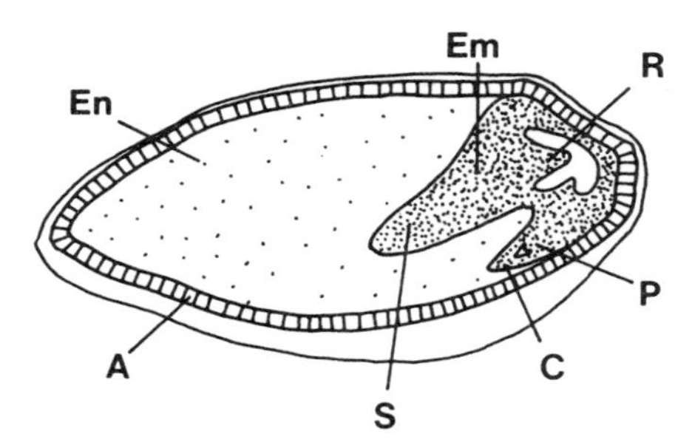

Fig. 8.9 Structure of the barley seed. *Key:* En = endosperm; Em = embryo; R = radicle; P = plumule; C = coleoptile; S = scutellum; A = aleurone layer.

'aleurone grains'). The aleurone cell walls are unusual in being composed mainly of arabinoxylan (85%), with only 8% of cellulose and 6% of protein, and are similar in structure to the walls of wheat endosperm cells. Early studies demonstrated that the starch of isolated barley endosperm would undergo hydrolysis if it was incubated in a culture flask with isolated embryos. Since the embryos themselves were shown not to produce amylases in significant amounts, it was concluded that they must produce a soluble factor or factors which induces amylase production in the endosperm tissue. It is now known that GA_3, one of the fifty or so naturally-occurring gibberellins, is the soluble factor, which interacts with cells of the aleurone layer to induce the synthesis of a specific group of enzymes including α-amylase, protease, ribonuclease, β-1,3-glucanase, phosphatase and endo-β(1→4)-xylanase. Only α-amylase and protease require the presence of GA_3 for their synthesis. These enzymes, together with others which are already present in the embryo or endosperm and are also not specifically induced by GA_3, such as β-amylase, catalyse the breakdown of both the stored starch and also of the walls and cytoplasm of the aleurone cells themselves. Ultrastructural studies have revealed that GA_3 induces the synthesis and stacking of the endoplasmic reticulum of the aleurone, and it is thought to be functionally related to enzyme synthesis. The mechanism of gibberellin action is not clear, but it has been demonstrated that the effect, which can be inhibited by abscisic acid, is at the level of transcription. Nucleotide sequences have been determined for the GA_3-inducible α-amylase and protease ('aleurain') genes (Whittier *et al.*, 1987). The 5′ region of the aleurain gene is rich in G + C sequences and is flanked with inverted sequences. Two sets of homologous sequences were also found in both genes, and a part of the conserved sequences are similar to the core sequence of certain enhancer elements found in mammalian cells. It is suggested that these conserved sets of sequences may play a role in determining either the tissue specificity of expression or in some interaction with GA_3 or abscisic acid.

β-Amylase, which constitutes 1–2% of the protein content of the barley endosperm, cleaves maltose from the non-reducing end of α(1→4)-D-glucans. Kreis *et al.* (1987) have isolated a cDNA and the mRNA for the enzyme, which is encoded by a small gene family, and determined the full nucleotide sequence.

Unlike α-amylase, the translation product of the mRNA appears to be the mature protein, lacking a N-terminal signal peptide, and it has been suggested that a sequence within the coding region of the gene might direct the intercellular transport of the enzyme, but this is by no means certain. The abundance of the α-amylase mRNA correlates well with the level of the mature protein found *in vivo* or when translated *in vitro*, suggesting that deposition of the enzyme is at least partly controlled at the transcriptional level.

The regulation of the mobilization of fats (triacylglycerols) to carbohydrate is less well understood at the molecular level, although the biochemical pathways, and the compartmentalization of the metabolic intermediates, are largely established. The breakdown of fats takes place in the storage tissues, and the sucrose which is ultimately formed is translocated to the developing embryo. The key biochemical pathway for this conversion is the glyoxylate cycle, the activity of which is essential to effect net gluconeogenesis from acetyl-CoA derived from the β-oxidation of the fats. Two enzymes, isocitrate lyase and malate synthase, are unique to this cycle, and are only found at high activity during germination of lipid-storing seeds. Extractable activities decline when the fats are mobilized. In the castor bean (*Ricinus communis*) mobilization begins on day 3 after imbibition, and is complete by day 7.

There is an interaction between three subcellular compartments during fat breakdown: the oil body (spherosome) which stores the lipid, the glyoxysome, and the mitochondrion (Fig. 8.10). The initial phase of lipid breakdown is the conversion of triglycerides to fatty acids and glycerol, catalysed by lipase (or phospholipase for phosopholipids), probably within the oil bodies. Glycerol can be converted to pyruvate or to sugars via glycerol phosphate and triose phosphate, while long-chain fatty acids are catabolized by the successive removal of 2-carbon fragments, in β-oxidation, the enzymes for which are localized in the glyoxysomes. It is also possible that fatty acids may be broken down by α-oxidation, in which they are peroxidatively decarboxylated to CO_2 plus a long-chain aldehyde, which is in turn oxidized to a corresponding acid. Although germinating seeds of peanut appear to possess the required enzymes, the extent to which this pathway contributes to energy metabolism *in vivo* is uncertain. What is known is that the acetyl-CoA produced by β-oxidation enters the glyoxylate cycle within the glyoxysome to yield one mole each of succinate and NADH from two moles of acetyl-CoA. The physical separation of the acetyl-CoA from the mitochondrial TCA cycle prevents its further oxidation. However, succinate is transferred to the mitochondrion where it is converted to oxaloacetate, phosphoenolpyruvate, and then to sugars.

The tissue-specific, and even organelle-specific, accumulation of isocitrate lyase and malate dehydrogenase at particular stages of seed development raises interesting questions concerning their regulation. Recently, cDNA sequences encoding the glyoxysomal malate synthase of cucumber (*Cucumis sativus*) and part of the sunflower (*Helianthus annuus*) seedling isocitrate lyase have been isolated (Smith and Leaver, 1986; Allen *et al.*, 1988), and have been used to study transcription of the genes during seedling development. During the first two days of germination and growth in the light, the rate of mRNA accumulation for both

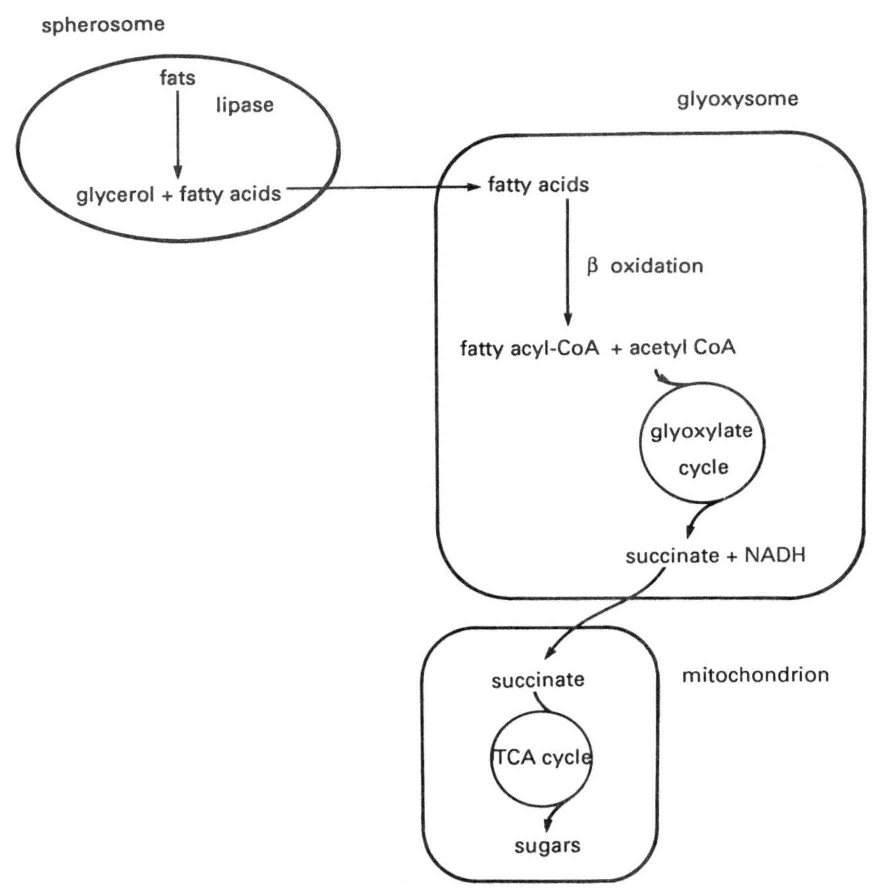

Fig. 8.10 The compartmentalization of the conversion of lipids to sugars in plant cells.

genes was greater than if the seeds were germinated in the dark, and the changes in relative transcript abundance correlated with changes in levels of the enzyme, as detected immunologically. These results indicate that the expression of the genes to be both developmentally and environmentally regulated, but putative regulatory sequences have yet to be identified.

Phytochrome

The observation that the germination of lettuce seeds is promoted by light of wavelength 660 nm, but is inhibited by a wavelength of 730 nm has been taken as evidence of the involvement of phytochrome. The subsequent growth of germinated seedlings is also strongly influenced by the availability of light, which

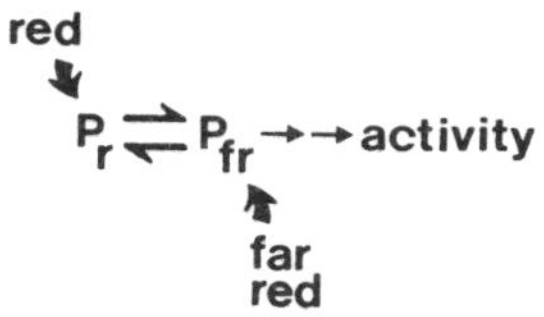

red

$P_r \rightleftarrows P_{fr} \rightarrow\rightarrow$ activity

far red

Fig. 8.11 Phytochrome interconversion. (See text.)

determines whether they will grow in an etiolated manner ('skotomorphogenesis') or as fully autotrophic shoots ('photomorphogenesis'). It is beyond the scope of this book to discuss in detail the biology of the phytochrome molecule, but because of its importance as one of the factors regulating seed and seedling development (amongst others) one or two points should be made. Recent reviews by Jordan *et al.* (1986) and Mohr (1987) cover the subject in great detail.

Phytochrome is a pigment system which operates predominantly in the red/far-red range of the spectrum. Other sensory pigments include 'cryptochrome' (operating in the blue/UV-A range) and a 'UV-B photoreceptor', which exhibits an action spectrum with a sharp peak at 290 nm. Biochemically, phytochrome is a chromoprotein comprising two distinct forms: P_r, which absorbs light of wavelength 660 nm, and in so doing is converted from a metabolically inactive state to P_{fr}, which is the active form ('r' = red; 'fr' = far red). P_{fr} absorbs at 730 nm, and is in turn converted back to inactive P_r (Fig. 8.11). In combination with the other sensory pigments, the plant is able to determine the quality of the available light, and modify its development or behaviour accordingly. It is this *plasticity* of development, augmentary to the genetic programme, which allows a rapid response to, and hence survival in, changing environmental conditions.

An example of the role of phytochrome, and its interaction with other light sensors, is described by Mohr (1987). The rate of elongation of the hypocotyl of the sesame seedling (*Sesamum indicum*) is controlled by white and blue light, while red light has no apparent effect on growth beyond 60 hours after sowing the seed. However, at between 36–60 hours after sowing, red light does have an inductive effect which can be reversed by far-red light, indicating that the responsiveness of the developing seedling changes. If, on the other hand, seedlings are maintained in blue light for 72 hours, the ability to respond to red/far-red light treatments is preserved, and this and other experimental data suggest that, while P_{fr} is the effector molecule, the ability of the plant to respond to red light is d etermined by events set in train by other, shorter wavelengths.

The phytochrome molecule has been the subject of molecular biological studies with a view to understanding and possibly modifying its regulation. Phytochrome has been isolated and purified from a range of tissues by both conventional and immunoprecipitation techniques, and is composed of two identical monomers, each of 124 kDA with a total of about 1100 amino acids. The abundance of phytochrome is strongly influenced both by the stage of development of the plant and also by the ambient light conditions. It is synthesized in the P_{fr} form, but interestingly, illumination in red or white light fo convert P_r to P_{fr} results in an 80–90% decline in the steady-state levels of phytochrome, an effect which can be reversed by far-red light; and this appears to be regulated at the level of transcription. Further studies of the regulation of an oat phytochrome gene have involved the construction of phytochrome-CAT translational gene fusions and the investigation of their expression in transgenic tobacco plants. When the plants were kept in the dark for 4 days following a 15-min far-red pulse, significant levels of CAT activity were detectable in extracts from the etiolated plants, but not in extracts of plants maintained in a normal 16 hour/8 hour dark regime. This is consistent with the effect of red (or white) and far-red light on transcript and phytochrome protein levels. The effects of various phytochrome constructions on the response to light are being studied, but results are not yet available.

General Reading

Bewley, J.D. and Black, M. (1978). *Physiology and Biochemistry of Seeds*, Vol. 1 and 2. Berlin, Heidelberg, New York, Springer-Verlag.

Bewley, J.D. and Black, M. (1985). *Seeds: Physiology of Development and Germination*. New York, London, Plenum Press.

Guern, J. (1987). 'Regulation from within: the hormone dilemma', *Ann. Bot.* **60**, Suppl. 4, pp. 75–102.

Knox, R.B. and Singh, M.B. (1987). 'New perspectives in pollen biology and fertilization', *Ann. Bot.* **60**, Suppl. 4, pp. 15–37.

Raghavan, V. (1986). *Embryogenesis in Angiosperms: a Developmental and Experimental Study*. Cambridge, Cambridge University Press.

Specific Reading

Allen, R.D., Trelease, R.N. and Thomas, T.L. (1988). 'Regulation of isocitrate lyase gene expression in sunflower', *Plant Physiol.* **86**, pp. 527–532.

Brady, C.J. (1987). 'Fruit ripening', *Ann. Rev. Plant Physiol.* **38**, pp. 155–178.

Chen, Z.-L., Pan, N.-S. and Beachy, R.N. (1988). 'A DNA sequence that confers seed-specific enhancement to a constitutive promoter', *EMBO J.* **7**, pp. 297–302.

Choi, J.H., Liu, L.-S., Borkird, C. and Sung, Z.R. (1987). 'Cloning of genes developmentally regulated during plant embryogenesis', *Proc. Natl. Acad. Sci. USA* **84**, pp. 1906–1910.

Christoffersen, R.E., Tucker, M.L. and Laties, G.G. (1984). 'Cellulase gene expression in ripening avocado, *Persea americana* cultivar 'Hass', fruit; the accumulation of messenger RNA and protein as demonstrated by complementary DNA hybridization and immunodetection', *Plant Mol. Biol.* **3**, pp. 385–392.

Colot, V., Robert, L.S., Kavanagh, T.A., Bevan, M.W. and Thompson, R.D. (1987). 'Localization of sequences in wheat endosperm protein genes which confer tissue-specific expression in tobacco', *EMBO J.* **6**, pp. 3559–3564.

Cyr, R.J., Bustos, M.M., Guiltinan, M.J. and Fosket, D.E. (1987). 'Developmental modulation of tubulin and mRNA levels during somatic embryogenesis in cultured carrot cells', *Planta* **171**, pp. 365–376.

Dewey, R.E., Diedow, J.N., Timothy, D.H. and Levings, C.S., III (1988). 'A 13-kilodalton maize mitochondrial protein in *E. coli* confers sensitivity to *Bipolaris maydis* toxin', *Science* **239**, pp. 293–295.

Dupuis, I., Roeckel, P., Matthys-Rochon, E. and Dumas, C. (1987). 'Procedure to isolate viable sperm cells from corn (*Zea mays* L.) pollen grains', *Plant Physiol.* **85**, pp. 876–878.

Dure, L., III (1985). 'Embryogenesis and gene expression during seed formation', *Oxf. Surv. Plant Mol. Cell Biol.* **2**, pp. 179–197.

Estrelle, M.A. and Somerville, C.R. (1986). 'The mutants of *Arabidopsis*', *Trends Genet.* **2**, pp. 89–93.

Jordan, B.R., Partis, M.D. and Thomas, B. (1986). 'The biology and molecular biology of phytochrome', *Oxf. Surv. Plant Mol. Cell Biol.* **3**, pp. 315–362.

Kreis, M., Williamson, M., Buxton, B., Pywell, J., Hejgaard, J. and Svendsen, I. (1987). 'Primary structure and differential expression of β-amylase in normal and mutant barleys', *Eur. J. Biochem.* **169**, pp. 517–525.

Leaver, C.J. and Gray, M.W. (1982). 'Mitochondrial genome organisation and expression in higher plants', *Ann. Rev. Plant Physiol.* **33**, pp. 373–402.

Marris, C., Gallois, P., Copley, J. and Kreis, M. (1988). 'The 5′ flanking region of a barley B hordein gene controls tissue and developmental specific CAT expression in tobacco plants', *Plant Mol. Biol.* **10**, pp. 359–366.

Meinke, D.W. (1986). 'Embryo-lethal mutants and the study of plant embryo development', *Oxf. Surv. Plant Mol. Cell Biol.* **3**, pp. 122–165.

Mishra, K.P., Joshua, D.C. and Bhatia, C.R. (1987). '*In vitro* electroporation of pollen', *Plant Science* **52**, pp. 135–139.

Mohr, H. (1987). 'Regulation from without: darkness and light', *Ann. Bot.* **60**, Suppl. 4, pp. 139–155.

Nasrallah, J.B. and Nasrallah, M.E. (1986). 'Molecular biology of self-incompatibility in plants', *Trends Genet.* **2**, pp. 239–244.

Quatrano, R.S. (1986). 'Regulation of gene expression by abscisic acid during angiosperm embryo development', *Oxf. Surv. Plant Mol. Cell Biol.* **3**, pp. 467–477.

Slater, A., Maunders, M.J., Edwards, K., Schuch, W. and Grierson, D. (1985). 'Isolation and characterisation of cDNA clones for tomato polygalacturonase and other ripening-related proteins', *Plant Mol. Biol.* **5**, pp. 137–147.

Smith, S.M. and Leaver, C.J. (1986). 'Glyoxysomal malate synthase of cucumber: molecular cloning of a cDNA and regulation of enzyme synthesis during germination', *Plant Physiol.* **81**, pp. 762–767.

Stinson, J.R., Eisenberg, A.J., Willing, R.P., Pe, M.E., Hanson, D.E. and Mascarenhas, J.P. (1987). 'Genes expressed in the male gametophyte of flowering plants and their isolation', *Plant Physiol.* **83**, pp. 442–447.

Tercé-Laforgue, T., Sallantin, M. and Pernollet, J.C. (1987). 'Wheat endosperm mRNA and polysomes and their *in vitro* translation products during development and early stages of germination', *Physiol. Plant.* **69**, pp. 105–112.

Whittier, R.F., Dean, D.A. and Rogers, J.C. (1987). 'Nucleotide sequence analysis of alpha-amylase and thiol protease genes that are hormonally regulated in barley aleurone cells', *Nucl. Acids Res.* **15**, pp. 2515–2535.

Chapter 9

Manipulation of Resistance

Plants are susceptible to the adverse effects of a wide range of external factors, some natural and some man-made, including pathogens, pests, chemical herbicides and extreme environmental conditions, such as high and low temperatures, drought and flooding. A number of strategies have evolved either to avoid or to overcome problems associated with the natural threats to survival, and mutants may appear in the field which are resistant to herbicides. As we discussed in Chapter 1, it has been the aim of plant breeders to transfer 'resistance genes' from otherwise valueless wild species into economically important crops, and it is hoped that molecular biological techniques will aid breeding programmes in two ways: (1) by contributing to our knowledge of the basis, and in particular the genetic basis, of the response of resistant and susceptible plants to a particular external factor; and (2) transferring the relevant genes to susceptible individuals. A list of some species which possess potentially useful genes is given in Table 9.1.

Resistance to Fungal and Bacterial Disease

Plants have available a range of defence strategies to counter attack by fungi and bacteria: (1) they may possess preformed chemical inhibitors to prevent proliferation of fungal mycelia; for example, the avenacins, found in oats, are glycosylated terpenes which act at concentrations of 3–50 ppm to prevent infection from *Gaeumannomyces graminis*; (2) they may produce secondary metabolites, termed 'phytoalexins' (meaning 'warding-off' compounds produced by plants) which are toxic and inhibit further invasion of the pathogen—

Table 9.1 A list of plants with some potentially useful resistance characters

Plant	*Tolerance*
Solanum brevidens	potato leaf roll virus
S. berthaultii	aphids
Chenopodiaceae	salinity
Cactaceae, Euphorbiaceae, *Larrea divaricata*	desiccation
rice, rushes, mangroves	flooding

phytoalexins are typically synthesized *de novo* on infection; (3) lignification of tissues may occur at the site of infection to create a physical barrier to hyphal penetration; and (4) rapid necrosis may occur at the site of infection to isolate the pathogen and prevent its further spread—this is characteristic of the so-called 'hypersensitive response'. The latter three strategies are of particular interest to biochemists and molecular biologists, because a defence response, which involves the differential expression of a number of specific genes, is triggered by the pathogen, and even by non-biological factors such as heavy metal ions. Such biological and non-biological triggers are generically termed 'elicitors'. Particular fractions of fungi, such as cell-wall fragments or components of culture filtrates, are known to act as elicitors, viable pathogens not being an absolute requirement to induce a response in the host (although viable rather than inactive bacteria do seem to be a requirement for the hypersensitive response to these microorganisms); however, the precise mechanism(s) involved have not been elucidated. Nevertheless, the change in the metabolism of the infected plant (or elicited cell culture) has attracted much attention both as a means of studying the levels at which specific genes are regulated, and as a means of identifying genes which might be involved in protection of the plant from disease. We will consider some examples of elicited gene expression to illustrate the general principles, as far as they are known, of defence against fungi.

Collinge and Slusarenko (1987) have described the strategies for investigating pathogen-elicited gene expression as falling into two complementary groups: first, the 'targeted approach', and second, the 'shotgun approach'. In the targeted approach, a particular biochemical event associated with the response to infection is characterized; defence-related proteins (e.g. enzymes in phytoalexin biosynthesis) are isolated and antibodies raised to them; corresponding transcripts can be identified following their translation *in vitro*; and cDNA clones can be made and used as probes to determine transcript levels corresponding to the genes encoding the proteins originally of specific interest. This approach assumes a functional significance of the elicited protein (the 'target'). In the shotgun approach, transcripts specifically elicited by infection are isolated (e.g. by differential hybridization of cDNA libraries prepared from elicited and non-elicited tissues) and the cDNA clones can be used to pick out genomic fragments encoding the genes. Here, there is no specific target to be isolated, merely a range of mRNAs of presumed regulatory importance, the precise nature of which may (or

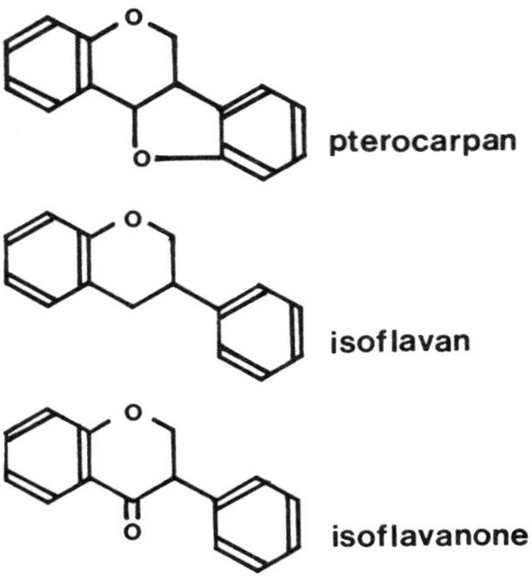

pterocarpan

isoflavan

isoflavanone

Fig. 9.1 Basic structures of pterocarpan, isoflavan and isoflavanone phytoalexins.

may not) be determined subsequently, in concert with studies using the targeted approach. These broad strategies have been described in Chapter 3, and have similarly been used to identify development- and stress-specific genes and gene products (see Chapter 8 and below). Some specific examples of differential gene expression in response to fungal attack are now briefly considered.

PHYTOALEXIN BIOSYNTHESIS

Most of the work on phytoalexin biosynthesis has been carried out in members of the Solanaceae and Leguminosae (reviewed by Dixon, 1986). The first such compound characterized was found to be a pterocarpan (Fig. 9.1), induced by the infection of pea (*Pisum sativum*) pods by the soft-fruit pathogen *Monilinia fructicola*, and was named 'pisatin'. Subsequently, the phytoalexin phaseollin, also a pterocarpan, was isolated from the French bean (*Phaseolus vulgaris*) infected with the same fungus. It is now known that the phytoalexins of the Leguminosae fall into three main groups, namely the pterocarpans, the isoflavans (e.g. vesitol and sativan found in *Lotus corniculatus*) and the isoflavanones (e.g. kievitone from *Phaseolus vulgaris* and wyerone and wyerone acid from *Vicia faba*). The first identification of phytoalexins in the Solanaceae was in 1968, which led to the characterization of the terpenoid rishitin, elicited in potato infected by *Phytophthora infestans*, and a large number of other inducible compounds, primarily terpenoids or terpenoid derivatives, have been identified in this family. Very few other families, however, have been studied for their phytoalexin response, and so firm evidence confirming the ubiquity of the reaction throughout the plant kingdom is not available. Nevertheless, isoflavones have been detected in response to infection in the Chenopodiaceae, isocoumarins in the Umbelliferae and naphthaldehydes in the Malvaceae.

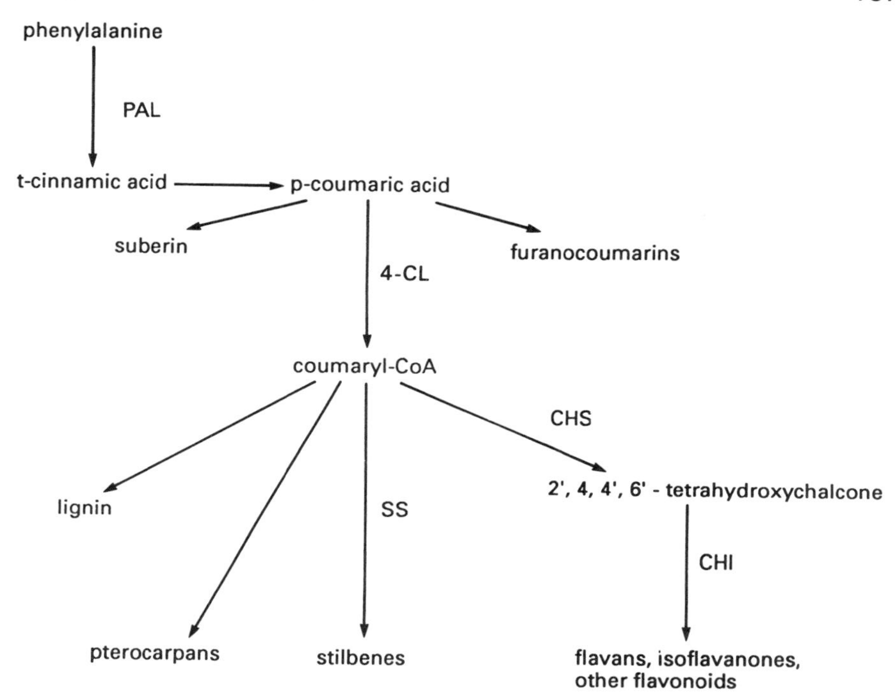

Fig. 9.2 Summary of phenylpropanoid metabolism.
Key: PAL = phenylalanine ammonialyase; CHS = 6′-hydroxychalcone synthase; CHI = chalcone isomerase; SS = stilbene synthase; 4-CL = 4-coumarate ligase.

The majority of biochemical and molecular evidence concerning the biosynthesis of phytoalexins has been obtained for the phenylpropanoid pathway, which is also involved in lignin synthesis (Fig. 9.2) and, to a lesser extent, for terpenoid metabolism. The basic flavonoid skeleton is a derivative of two converging pathways, the acetate–mevalonate pathway and the shikimic acid pathway. Phenylalanine, which is produced from shikimic acid, is deaminated to cinnamic acid by the enzyme phenylalanine ammonia lyase (PAL). Cinnamic acid is then hydroxylated to *p*-coumaric acid (4-hydroxycinnamic acid) which can enter the phytoalexin pathways by esterification with acetyl CoA (via 4-coumarate: CoA ligase; 4-CL) to form chalcones (via chalcone synthase; CHS), isomeric flavanones (via chalcone isomerase; CHI) and stilbenes (via stilbene synthase) (Fig. 9.2). Most of the subsequent metabolic steps have not been clearly defined. Terpenoids such as rishitin appear to be derived from acetate and mevalonate, but again, the details of the terminal parts of the pathway, and control points, are not clear. The regulation of the phenylpropanoid pathway is better understood, and has been investigated by studying changes in the activity of a relatively small number of enzymes, and more recently of their encoding genes, under conditions of elicitation, commonly in cell cultures of parsley

(*Petroselinum crispum*) and French bean (*Phaseolus vulgaris*). For example, cultures of both species exhibit transient increases in the activities of PAL, 4-CL, CHS, CHI and at least one cinnamate hydrolyase when treated with fungal preparations; and for PAL, CHS and CHI these increased activities are correlated with increased rates of synthesis of the enzymes *in vivo*. Transcriptional studies have further shown a correlation between increased enzyme synthesis and the increased abundance of enzyme-specific transcripts. The response to elicitor treatment is very rapid, with transcripts of PAL and CHS appearing within 5 min. in bean; CHI transcription and enzyme activity may in some systems appear slightly later, but for all three enzymes transcript abundance usually peaks at 3–4 hours after elicitation (Dixon *et al.*, 1986; Mehdy and Lamb, 1987). Stilbene synthase mRNA is also rapidly induced in elicited peanut cells. Similar correlations between the induced activities of PAL and 4-CL and the abundance and translational activities of the corresponding mRNAs have been observed in potato leaves and cell cultures elicited with *Phytophthora infestans* or preparations of it. The functional significance of these increases has also been investigated by infecting leaves of susceptible and resistant potato cultivars with virulent and avirulent races of the fungus. The initial transient increases in abundance of both PAL and 4-CL mRNAs are almost identical in compatible (disease-forming) and incompatible (no disease response) interactions, and are unlikely to play a significant role in race-specific resistance. Since this period of gene activation occurs during the period of hyphal penetration of the tissue, followed by browning of the tissue, it is presumably associated with damage-related phenolics production, perhaps as part of the hypersensitive response. However, in compatible reactions, the early peak of mRNA induction is followed after approximately 7–12 hours post-infection by a second increase in PAL and 4-CL mRNA levels, which is not observed in the incompatible 'infection'; this reaction may be related to hyphal penetration of the opposite epidermis.

PAL and CHS exist, respectively, as isozymic forms, and are encoded by multigene families. Elicited bean cells, however, only accumulate a single polypeptide of CHI, and evidence from restriction site analysis and genomic blotting indicates that this enzyme is encoded by a single gene only (Mehdy and Lamb, 1987). Work is in progress to isolate full length genes of these inducible enzymes, to determine *cis*-acting regulatory sequences which may determine that inducibility, but no results have as yet been published.

CELL-WALL MODIFICATION

Lignification provides a major structural barrier to hyphal invasion of tissues, notably in grass species. The early intermediates of phenylpropanoid metabolism are common to the biosynthesis of both flavonoid phytoalexins and lignin (Fig. 9.2), but other enzymes such as peroxidase, *p*-coumarate:CoA ligase and ferulate:CoA ligase activities are also elicited in relation to this defence mechanism. 1-Aminocyclopropane-1-carboxylic acid (ACC) synthase, a key enzyme in ethylene biosynthesis (see Chapter 8) is similarly induced by fungal infection, and ethylene in turn induces large increases in mRNA levels of PAL, 4-

CL, CHS and also of hydroxyproline-rich glycoproteins, the syntheses of which appear also to be involved in cell-wall modification on infection.

Callose, a β-(1→3)-glucan, is a component of 'papillae', cell-wall invaginations at the site of hyphal invasion, but its increased accumulation appears to be the result of the activation of β-(1→3)-glucan synthase by calcium ions, rather than by *de novo* synthesis.

HYDROLYTIC ENZYMES AND PR PROTEINS

Apart from enzymes associated with secondary and cell-wall metabolism, a number of other proteins are elicited by pathogen attack and have been grouped together as 'pathogenesis-related' or 'PR' proteins. A similar range of polypeptides may appear in plant tissues infected with fungi, bacteria or viruses, or treated with heavy metal ions or UV light (see below). PR proteins are usually defined as polypeptides of relatively low molecular weights (10–40 kDa) which accumulate extracellularly in infected tissue, and generally exhibit high resistances to proteolysis and commonly have extreme isoelectric points. The specific functions of these proteins are largely undetermined, but there is evidence that some are directly involved in defence mechanisms against fungi. Some PR proteins, elicited in potato leaves infected with *Phytophthora infestans*, have been identified recently as β-(1→3)-glucanases and chitinases. Both types of enzyme are known to cleave carbohydrate fractions from fungal cell walls, which commonly contain chitin (poly[1,4-(*N*-acetylglucosamine)]) and chitosan (deacetylated chitin), so inhibiting growth; and endochitinase also possesses lysozyme activity, and so may be protective against pathogenic bacteria.

The regulation of chitinase synthesis appears to vary between species. In French bean infected with *Pseudomonas syringae* pv. *phaseolicola*, the appearance of translatable chitinase mRNAs is detectable as early as 6 hours after inoculation with an avirulent race of the pathogen, but only after 24 hours if infection is with a virulent strain. These results indicate that the inhibition of disease formation is associated with the early, specific expression of the chitinase gene. In this species, chitinase is encoded by a multigene family, and the transcription of at least two of the members is induced by ethylene. The ethylene dependence is not, however, observed for elicited pea.

HYPERSENSITIVE RESPONSE

The hypersensitive response to infection is characterized by a series of physiological changes, including browning, callose formation, vascular blockage and phytoalexin accumulation, leading to necrosis of the tissues. Although it is associated with the attenuation of the spread of disease, the precise role of cell death is unclear. It has been suggested that necrosis itself prevents the progress of infection, but it is also possible that tissue death is a secondary effect, a visual marker that the hypersensitive response has occurred, and the early events are perhaps more important than necrosis *per se*. There is evidence of differential gene expression during the early phase of the response before any visible changes in the tissue are apparent. For example, necrosis of bean leaves inoculated with avirulent *Pseudomonas syringae* pv. *phaseolicola* occurs after 21–25 hours, but this is preceded by changes in the steady-state levels of PAL mRNA, which begin to increase between 6 and 9 hours after infection, and peak at about 12 hours. In a study of the hypersensitive response by *Brassica campestris* to *Xanthomonas campestris*, Collinge *et al.* (1987) found that while the virulent pathovar *X.c.* pv. *campestris* induced chlorosis and rotting of the leaves after 48 hours post-inoculation, the avirulent *X.c.* pv. *vitians* elicited a necrosis within about 12 hours. Differential gene expression was demonstrated by the *in vitro* translation of poly(A^+)RNA isolated over 24 hours after inoculation. A number of transcripts appear within 4 hours which were specific to the incompatible response, and translated into polypeptides and, unlike PR proteins, exhibited a wide range of size and isoelectric points. Their role in defence, however, was not elucidated.

We have given an outline of the principal defence strategies of plants to attack by fungi and bacteria, an understanding of which is an essential first step to genetically manipulating susceptible genotypes. We are beginning to identify microbial factors which elicit activity in plant tissues, leading to an incompatible response, and *in vitro* studies have demonstrated that a number of the compounds elicited, notably phytoalexins and chitinases, successfully inhibit pathogen growth. Recent evidence suggests that some PR proteins may act in the same way. While the transfer of chitinase genes to susceptible plants would seem to be quite possible, the transfer of complex secondary metabolic pathways is far less straightforward, and much more basic biochemistry must be carried out in this area (see Chapter 7). The mechanisms of the recognition of potentially pathogenic microorganisms by plants is an area which requires more study (see Callow *et al.*, 1987). We have seen how, in compatible interactions, there is a rapid response to infection, and while the subsequent events of the response may be characterized, the genetic control of the recognition event itself is poorly understood. Models have been proposed which attempt to explain recognition in terms of the interaction of molecules of pathogen origin with receptors borne by the plant, perhaps encoded by resistance genes (Lutgenberg, 1986). Recent evidence suggests that pathogens may be able to evade recognition by host plants, evolving mechanisms to overcome the defence systems. Kearney *et al.* (1988) have shown how *Xanthomonas campestris* pv. *vesicatoria*, which causes bacterial spot disease of peppers (*Capsicum annuum*), mutates to bypass the genetically-defined recognition system by the transposon-induced alteration of a specific gene, $avrBs_1$. This gene is known to specifically induce a hypersensitive response in pepper plants carrying a resistance gene, Bs_1. Whether transposon mutagenesis is a universal mechanism by which pathogens can overcome resistance is not yet known.

It may be possible to select for plant resistance genes *in vitro*. For example, Shahin and Spivey (1986) have isolated cultured cells of tomato which are resistant to fusaric acid, a toxin produced by the fungus *Fusarium oxysporum*, and regenerated plants were produced which possessed a single dominant gene encoding resistance to *Fusarium* wilt. The cloning of this and of other single dominant resistance genes, and the study of their expression in transgenic plants, is an exciting prospect.

BIOLOGICAL CONTROL

A quite different approach to controlling disease caused by fungi and bacteria derives from the observation that some species of bacteria release metabolites which inhibit the growth of other pathogenic species (see review by Davison, 1988). For example, the common soil bacterium *Pseudomonas fluorescens* produces both anti-fungal antibiotics and 'siderophores'. Siderophores are molecules which chelate Fe^{3+} ions with high affinity, and act as an iron-scavenging system. It is thought that the depletion of the local environment of iron inhibits the growth of other microorganisms, notably pathogenic fungi such as *Gaeumannomyces graminis* (causing 'take-all' of cereals) and *Fusarium oxysporum* (causing wilt diseases), and bacteria such as *Erwinia carotovora* (causing soft rots) and *Pseudomonas syringae* (causing Halo blight and wildfire disease). In field trials, wheat yields have been increased 27% by inoculation of seeds with *P. fluorescens*, and similar treatment of potato seed tubers has been found to increase tuber yields by as much as 70%.

A method has also been described for the biological control of crowngall disease, which causes major economic losses in fruit trees and ornamentals. As described in Chapter 6, this disease is caused by *Agrobacterium tumefaciens*. The prevention of disease formation is achieved by spraying young plants or seeds with the avirulent species *Agrobacterium radiobacter* var. *radiobacter* which produces a plasmid-encoded antibiotic, agrocin 84, inhibitory to nopaline strains of *A. tumefaciens*; and selection of strains which produce agrocin effective against octopine and agropine types have now been cloned, and have been transferred to the root symbiotic species *Rhizobium meliloti*. The ability of the modified bacterium to control root tumours is not yet known. It is also possible that genetic engineering techniques could be used to improve further these beneficial bacteria by manipulating such properties as antibiotic/siderophore yield, chemotactic responses and root adhesion, or improved survival under adverse environmental conditions.

Resistance to Viruses

Viral diseases of crop plants constitute a major economic problem through reduction in product yield, and virus biology has been studied for many years (reviewed by Zaitlin and Hull, 1987). The genetic material of viruses may be either DNA or RNA, and may be single- or double-stranded. Approximately 77% of characterized plant viruses possess a single plus-(messenger) sense strand of RNA. Infection of plant tissue requires damage to the cell wall and/or plasma membrane, which, for insect-borne viruses, is achieved by the penetration by the insect stylet during feeding. Once inside the cell, the virus particle is uncoated to release its nucleic acid, and for at least some plus-stranded RNA viruses, such as tobacco mosaic virus (TMV), uncoating is achieved by cytoplasmic ribosomes which also translate the RNA. Plant virus nucleic acids are not integrated into the host genome. Common translational products amongst most, if not all, viruses, include (1) coat protein, (2) one or more proteins involved in the replication process, and (3) factors involved in the systemic transmission of the virus away from the site of infection. A variety of strategies have evolved by which viruses produce the several proteins required. For example, cowpea mosaic virus, which has two RNAs, translates both completely to produce two separate translational fusions (polyproteins), which are subsequently cleaved to produce the mature polypeptides. Tobacco mosaic virus separates its genomic RNA into monocistronic fragments, probably during a transcription process, each of which are translated separately. Turnip yellow mosaic virus employs a combination of these strategies, and variations exist for other viruses (see Zaitlin and Hull, 1987).

Since eukaryotic cells synthesize RNA from DNA and genomic replication is restricted to the S phase of the cell cycle, RNA viruses encode the proteins comprising a RNA-dependent RNA polymerase (RdRp) complex for their own replication. The caulimoviruses, as typified by the cauliflower mosaic virus, have double-stranded DNA as their nucleic acid, and replicate in a biphasic process. The viral DNA is transcribed by the plant-encoded DNA-dependent RNA polymerase-II, and the RNA produced acts as a template for transcription back to DNA using a virus-encoded reverse transcriptase. After replication of the nucleic acids, it is packaged into newly synthesized coat protein to produce daughter virus particles. The only other group of plant DNA viruses, the gemini viruses, contain single-stranded DNA which is thought to be replicated via a double-stranded intermediate.

Following replication, the successful establishment of the infection requires the dissemination of the virus through the plant. In incompatible plant–pathogen interactions, a hypersensitive response may be elicited, and the spread of the particles is prevented. PR proteins may also be synthesized by the host. If, however, the plant is susceptible, the systemic movement of the virus may occur. Electron microscopical evidence indicates that virus particles are transmitted from cell to cell symplastically, i.e. via the cytoplasmic connections (plasmodesmata) which exist between adjacent cells. In a sense, the plant contains a single, continuous cytoplasm, and, with perhaps the exception of the very tip of the shoot meristem (see Chapter 4), is accessible to virus particles (although infection of seed and pollen is rare). There is evidence that the transmissibility of the virus is genetically encoded. A tomato strain of tobacco mosaic virus (LS-1) has been isolated which, at elevated temperatures, is unable to move through the plant. *In vitro* studies in protoplasts have shown that the replication *per se* of the virus is not temperature-sensitive, and a specific 30 kDa protein has been implicated as determining transmissibility. Sequence analysis has shown that the gene for the protein from LS-1 differs from that of the temperature-resistant strain by a single base-pair mutation, resulting in the protein possessing a serine instead of a proline. Transfer of the wild-type gene to transgenic plants allows the LS-1 virus to move through those plants at the restrictive temperature, demonstrating the virus' requirement for this protein (R.N. Beachy and co-workers, reported in Zaitlin and Hull, 1987). Proteins of similar function have been identified in other viruses, although their mode of action is unknown.

Although the symptoms of viral infection, such as the hypersensitive response, chlorosis and PR protein synthesis, may be readily detectable, the mechanisms by which the virus effects these physiological perturbations is largely undetermined.

Many virus diseases are named after the phenotypic changes which occur in leaves of the infected plants (tobacco mosaic virus; cowpea chlorotic mottle virus; the sugar-beet 'yellows'), and this effect appears to be due to the inhibition of chloroplast rRNA and protein synthesis, leading to a reduced abundance of chloroplasts. However, these effects are not necessarily correlated simply with virus replication, and presumably viral gene products interact with the plant cell in a very complex way—much more work needs to be done to define the viral genes responsible, before the effects can be specifically inhibited by genetic engineering techniques. However, some progress is being made in this respect. Plants which are systemically infected with cauliflower mosaic virus (CaMV) exhibit a characteristic mosaic patterning of the leaves, the cells of which accumulate viral inclusion bodies in the cytoplasm. The major protein of these inclusion bodies, P_{66}, is encoded by gene VI of CaMV, translated from the 19S RNA, one of the two transcripts produced by this double-stranded DNA virus. Gene VI has been implicated in host range control and in symptom production, and Baughman *et al.* (1988) have produced further evidence to support this view. A segment of the viral genome carrying gene VI was introduced into tobacco plants by *Agrobacterium*-mediated gene transfer, and the transgenic plants were found to display viral-like symptoms; deletions and frame-shift mutations of gene VI prevented the formation of disease symptoms in transgenic plants. There was also a direct correlation between the appearance of symptoms and of the amount of P_{66} accumulated by the plant, as detected by immunoblotting.

We have seen in Chapter 4 how viruses can be eliminated from individual plants by meristem culture, but traditional methods to prevent the spread of viral disease generally rely on the prevention of infection. This can be achieved by the use of resistant cultivars, but such an approach is limited by the size of the available gene pool. As is the case with fungal and bacterial diseases, the mechanisms of resistance, and particularly the recognition systems involved, are not understood in any detail. An alternative approach to limiting virus infection has involved the deliberate infection of susceptible crop plants with a mild strain of a particular virus (which does not induce severe symptoms), for it has been shown that, on subsequent infection of the same plant material with a strongly virulent strain of the same or a closely related virus, disease symptoms may fail to develop. Such a phenomenon, superficially analogous to the vaccination of animals, is termed 'cross-protection', and has been successfully employed for the control of viral diseases of, for example, tobacco, *Citrus* spp. and *Papaya* spp. There are, however, problems with this technique: the mild strain of virus may mutate to a virulent form; it may act synergistically with other viruses present to adversely affect plant growth and development; a virus which is mild for one species may have virulent effects on a second crop, and so cross-contamination may be a problem; and mild strains of virus, while not inducing severe symptoms on a crop, may nevertheless reduce product yield to some extent.

A knowledge of the molecular biology of aspects of virus function has led to the proposal of three general strategies for plant protection against viruses using genetic engineering techniques: (1) modified cross-protection; (2) the use of satellite nucleic acids; and (3) the use of anti-sense RNA.

The rationale of a modified cross-protection technique is to identify the viral genes and gene products which are responsible for the observed protection phenomenon, and to separate them from genes encoding the proteins responsible for symptom formation. Transgenic plants can then be generated which possess the protective genes, and would therefore be expected to be resistant to virulent strains without the dangers associated with introducing complete virus particles.

This technique has been used with some success. For example, Abel *et al.* (1986) have isolated the tobacco mosaic virus (TMV) coat protein gene and introduced it into tobacco plants under the control of the CaMV 35S promoter. In those plants in which the gene was expressed, symptoms developed more slowly than in wild-type plants following inoculation with TMV, and 10–60% of the plants producing the gene product developed no disease symptoms at all. Loesch-Fries *et al.* (1988) have demonstrated that tobacco plants transformed with the coat protein gene of alfalfa mosaic virus (AMV) are similarly protected against the full virus disease, and it was found that protection was specifically against AMV virus particles—both AMV RNA and TMV were as infectious in plants expressing the coat protein as those plants which did not express it.

The mechanism of cross-protection is unclear, but these results demonstrate specifically the importance of the coat protein. It has been suggested that cross-protection may confer resistance by preventing the uncoating of other invading viruses, a process which is an essential first step for nucleic acid replication and protein synthesis. This would account for the susceptibility of cross-protected plants to naked viral RNA, which obviously is already 'uncoated'. Perhaps the coat protein produced in transgenic plants competes for factors required by invading viruses for the uncoating process. Other proteins, such as the 30 kDa 'movement' protein of TMV, do not confer cross-protection when expressed in transgenic plants. It was observed by both Abel *et al.* (1986) and Loesch-Fries *et al.* (1988) that protection against symptoms was not absolute, and this may be related to the relatively low levels of coat proteins produced by the transgenic plants—less than about 1% of the coat protein found in virus-infected plants, in the latter report. This suggests that the concentration of coat protein may become limiting and determine, to some extent at least, the protection obtainable. The use of stronger gene promoters than the CaMV 19S (used in the AMV work) might produce higher levels of resistance.

A second approach for protection against viruses makes use of the observation that, in some plant RNA viruses, certain RNA sequences may ameliorate disease symptoms, in infected plants. These small sequences, which are replicated and packaged normally, but bear no sequence homology to the main genomic RNA and are not required for virus replication and spread, are termed 'satellite RNAs'. Their replication and transmission is dependent on factors encoded by the viral genome, and they are presumably of strong selective advantage to the virus since they appear not to be lost either during virus proliferation or through natural selection. An alternative view might be to think of satellite RNAs as being parasitic on the virus. The strategy for genetically engineering protection is to introduce DNA copies of satellite RNA into susceptible plants, which, on transcription, might be expected to inhibit symptom formation.

Harrison *et al.* (1987) have transformed tobacco plants with a DNA copy of a satellite RNA of cucumber mosaic virus (CMV) under the control of the CaMV 35S RNA promoter, using *Agrobacterium* as a vector. It was found that transformed, but non-infected plants, produced only low levels of satellite RNA, but on infection with satellite-free CMV the synthesis of large amounts of the RNA was induced. Although the 'background' amount of transcribed satellite RNA varied several-fold in non-infected transformants, similarly large levels accumulated in transformants infected with CMV. The accumulation of satellite RNA was correlated with a reduction in symptom formation. On infection with CMV, both control (untransformed) and transformed plants developed chlorotic lesions on the inoculated leaves. However, whereas control plants developed mosaic symptoms in all systemically invaded leaves and were stunted in their growth habit, transformed plants showed mosaic symptoms only in the first two or three leaves to be infected systemically, and leaves which were produced subsequently showed no symptoms.

The replication of CMV in the transformed plants was assayed by three methods: (1) Northern blot analysis, using both viral genomic RNA and satellite RNA as probes to determine the effect of satellite RNA on replication of the genome; (2) immunoassay of the abundance of the viral coat protein; and (3) the infectivity of cell extracts was tested on *Chenopodium amaranticolor*, a host of CMV which forms local lesions in proportion to the titre of biologically-active virus particles. It was found that the reduction in symptoms was correlated with the decreased replication of the virus, as indicated by all three tests, and these effects of satellite RNA were only apparent in systematically invaded leaves. To determine the viral specificity of these effects, transformed tobacco plants were inoculated with a range of taxonomically diverse viruses. Only tomato aspermy virus (TAV), which is very closely related to CMV, induced the synthesis of CMV satellite RNA, and disease symptoms were correspondingly reduced. Interestingly, however, the symptom attenuation was not correlated with either reduced TAV genomic RNA replication or with reduced infectivity of tissue extracts, in contrast with the effects on CMV proliferation.

Using a similar approach, Gerlach *et al.* (1987) investigated the effects of tobacco ringspot virus (TobRV) satellite RNA on symptoms produced by that virus in transgenic tobacco plants. In this system it was also observed that transcription of the introduced cDNA equivalent of the satellite RNA was induced dramatically by infection with virus, provided that the cDNA comprised a trimer (i.e. three copies in tandem) of the monomeric satellite. If only a monomeric cDNA was integrated, the transgenic plants exhibited no large increase in satellite RNA, indicating an effect of cDNA copy number, also observed by Harrison *et al.* (1987). The trimeric TobRV cDNA produced only monomer-sized transcripts, suggesting that cleavage of trimeric transcripts may have occurred in the transgenic plants. Furthermore, a reduction of ringspot symptoms was obtained in the trimer-, but not in the monomer-, satellite-transformed plants, and protection was correlated with a decreased level of replication of infectious virus particles.

The mechanism of satellite RNA action in its protective role is unknown. The observation that TAV symptoms may be reduced without necessarily completely inhibiting viral replication is somewhat difficult to explain, but Harrison *et al.* (1987) tentatively suggest the possibility that the CMV satellite RNA may have some effect on the early stages of TAV RNA replication, and symptom formation was dependent not so much on the final amount of viral RNA accumulated, but perhaps on the timing of its accumulation in relation to cell development—a retardation in replication may account for reduced symptoms. It may also be possible that the satellite RNA, or one or more of the proteins encoded by it, directly interacts in an as yet uncharacterized way with the genomic RNA or the symptom-producing process.

An interesting feature of the experiments described is that, even though the cDNAs corresponding to the satellite RNAs were under the control of the CaMV 35S RNA promoter, which is considered to be constitutive, transcription was amplified to a large extent only when the transgenic plants were inoculated with infective virus—the satellite was dependent upon the virus for replication. The protection was therefore 'induced' to meet the needs of the plant, and this contrasts with the coat-protein protection mechanism, which can be saturated by excess invading virus. A problem with the satellite RNA approach is that the sequences which are protective in one species can be virulent in a different species, or can mutate to a form which is virulent in the originally protected species. Future work will involve both detailed functional analyses of satellite RNAs, and attempts to modify their transmissibility.

A third general approach to inhibit viral infection of plants involves the possibility of using anti-sense RNA. As described in Chapter 3, anti-sense RNA (minus-strand RNA) binds to sense (plus-strand or messenger) RNA to prevent its translation, and virus replication, packaging and/or systemic transmission could conceivably be inhibited in transgenic plants which encode an anti-sense strand to a specific viral sequence. Although there is evidence that anti-sense RNA may inhibit, for example, the synthesis of specific heat-shock proteins (Chapter 3), there is as yet no decisive data on its use to reduce viral symptoms. A number of questions have to be answered; how much anti-sense RNA is required to ensure protection?; which sequences in the viral genome are the best targets for inhibition of viral function?; what is the minimum length of anti-sense RNA to effectively inhibit a specific gene?; and so on.

Very recent studies have demonstrated that some types of RNA, including the satellite RNA of tobacco ringspot virus, are capable of spontaneous and specific cleavage. It appears that this autolytic reaction is related to particular short sequences in the RNA, which confer a characteristic 'T-shape' or 'hammerhead' secondary structure. In the laboratory of W. Gerlach and J. Haseloff (unpublished), artificial 'ribozyme' molecules have been created which can cleave target mRNAs (e.g. that of the marker gene chloramphenicol acetyltransferase) and so prevent full expression of the encoding gene. The ribozyme molecule comprises two 8-nucleotide sequences which are homologous to, and so allowing hybridization sequences which are homologous to, and so allowing hybridization to, regions of the target RNA which flank the sensitive autolytic site—cleavage occurs at 5'-GUX-3', where X is an unpaired nucleotide. Like the use of anti-sense

RNA, the potential for such a technique is enormous, both in virology and in the study of other aspects of metabolic and developmental gene regulation.

VIRUS TESTING

An understanding of the molecular biology of viruses has greatly improved the opportunities for the early detection of the viral infection of plants. Antibodies raised to the coat protein provide an immunological method for the quantitative or semi-quantitative determination of virus particles in purified or even crude plant extracts. Most commonly, antibodies are linked to fluorescent markers, such as fluorescein isothiocyanate (FITC), or to enzymes, such as alkaline phosphatase (enzyme-linked immunosorbent assays, ELISA). Viroids, however, are different from viruses in that, although they may cause virus-like diseases, they are simply RNA molecules, usually 200–400 nucleotides in size, and lack a coat protein; they themselves therefore cannot be detected by immunological techniques, and it is also believed that their RNA is not translated, so it is not possible to trace them by looking for antigenic gene products. Instead, it has been possible to use nucleic acid hybridization techniques, such as RNA dot blots, to detect the viroid 'genome' in plant extracts using viroid RNA to prepare a radioactive probe. The technique is both highly specific and sensitive, and nitrocellulose or nylon filters on which the extracted nucleic acid is bound can be re-used, after washing, to test for the presence of a number of different viroids or viruses.

Resistance to Insects

Insects and nematodes may reduce the yield of a crop both by acting as vectors for viral disease and by causing mechanical damage to both aerial and underground tissues. The response to mechanical damage of plants is particularly intriguing, since it is not confined to the wound site itself, but is to some degree systemic in nature—biochemical activity characteristic of wounding is detectable at a distance from the site of damage. The physiological changes in insect-damaged tissues bear some similarity to the reaction to fungal infection, notably the deposition in the cell walls of lignin and suberin. A range of proteins accumulate, and include: PAL and peroxidases, associated with lignification and fatty acid synthetase, in potato tubers; extensin, an hydroxyproline-rich glycoprotein, in carrot tap roots; and proteinase inhibitors in the leaves of the Solanaceae and Leguminosae. Other wound-induced changes include the onset of cell division and the associated events required for energy transduction, described in Chapter 2 in relation to callus formation.

The proteinase inhibitor proteins and their genes have received much attention. When leaves of, for example, potato or tomato are damaged by insect attack or by other forms of mechanical damage such as crushing (which perhaps mimicks the chewing action of insects), serine proteinase inhibitor proteins accumulate both in the damaged leaves (the 'local' reaction) and also in undamaged leaves at a distance from the original wound site (the 'systemic'

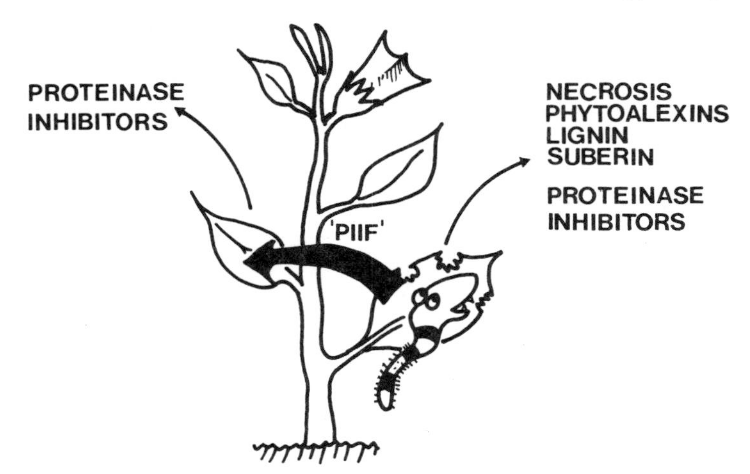

Fig. 9.3 Wound response in Solanaceous species. On wounding the systemic accumulation of proteinase inhibitors is induced, mediated by a 'proteinase inhibitor inducing factor' ('PIIF').

reaction). In non-wounded plants these proteins are generally restricted to storage tissues, such as seeds (e.g. the chymotrypsin inhibitors of barley) or tubers (e.g. patatin of potato), and in this situation their physiological role is not clearly understood. However, they are believed to play a role in defence against insects by virtue of their inhibitory action against proteases of insect, but not plant, origin and may interfere with the insect digestive process.

The systemic reaction requires that a signal be transmitted from the site of local damage to the more distant leaves (Fig. 9.3), and such a signal or 'wound hormone' has been called the 'proteinase inhibitor inducing factor' or PIIF. It has been demonstrated that pectinaceous plant cell wall fragments, or 'oligosaccharins', can influence various aspects of plant metabolism and development, and fragments released by endopolygalacturonase activity can induce the accumulation of proteinase inhibitors. It is possible, therefore, that oligosaccharins released by mechanical damage may act as the signal to induce the systemic reaction, although the transmissibility of such fragments through plant tissues has yet to be demonstrated.

The expression of the potato proteinase inhibitor-II (PI-II) gene has been studied in some detail. In the non-wounded potato plant the protein is localized to the tuber, but wounding of leaf tissue leads to its synthesis and accumulation in aerial parts, and both cDNA and genomic clones of PI-II have been isolated. Sanchez-Serrano *et al.* (1987) have used *Agrobacterium tumefaciens* to transfer the PI-II gene to tobacco plants, which do not normally possess homologous DNA sequences, to study aspects of its regulation in transgenic plants. Using a cDNA of the gene as a probe for PI-II mRNA, it was found that, whereas little or no expression of the gene was detected in non-wounded leaves, high levels of the mRNA were detectable in wounded leaves. The level of expression was as high in some of the transgenic tobacco plants as in wounded leaves of potato plants, and the gene was transcribed from the same initiation site in both species. This indicated that, not only the general regulatory mechanism, but also mRNA processing, was the same in both species. Furthermore, the systemic induction of the gene was demonstrated in the tobacco plants, and induction could also be achieved using the 'artificial' oligosaccharins chitosan and polygalacturonic acid instead of wound-induced damage. Thornberg *et al.* (1987) have also used transgenic tobacco plants to study the regulation of a PI-II gene. A 1000-base pair restriction fragment of the 5′ flanking region of the gene was linked to the chloramphenicol acetyltransferase (CAT) coding region, and to one of two terminal sequences: (a) that from the PI-II gene or (b) that from a Ti plasmid gene. It was found that the 5′ flanking region of PI-II possessed sequences which determined both the local and systemic expression of the CAT gene in response to mechanical damage, but only if the PI-II terminator was present—this result shows that the 3′ flanking region, as well as the 5′ region, is essential in determining the tissue- and environmentally-induced regulation of the gene. Interestingly, the 3′ region of the PI-II bears some homology to that of the extensin gene, which is also induced by damage. The time-course of CAT expression after wounding was also similar to the time-course of proteinase inhibitor accumulation in wounded tomato plants.

These experiments suggest two possible approaches for the engineering of resistance to insects. First, it is possible to transfer proteinase inhibitor genes to plants which do not normally synthesize them, and their expression has been shown to be inducible. This begs the question, do these proteins really play a protective role? Plant transformation techniques provide a direct way of testing this, and one significant example is described below. Second, the identification of *cis*-acting regulatory sequences of wound-inducible genes may prove valuable for engineering the *inducible* expression of genes encoding insecticidal products. Some possibilities for such genes are also considered.

Direct evidence for the effectiveness of proteinase inhibitors in reducing insect damage has been obtained by Hilder *et al.* (1987). A gene encoding a trypsin inhibitor of cowpea (*Vigna unguiculata*) was transferred, via *Agrobacterium tumefaciens*, into tobacco plants, and synthesis of the protein in the transformed plant was confirmed immunologically by Western blotting. The cowpea trypsin inhibitors had previously been correlated with resistance of cowpeas to the bruchid beetle *Callosobruchus maculatus*, and feeding experiments demonstrated their toxicity to a range of insect genera, including *Heliothis*, *Spodoptera*, *Diabrotica* and *Tribolium*, all of which cause losses of economic significance; there is no evidence of the toxicity of these proteins to either rats or humans. To determine any introduced resistance in the transgenic tobacco plants, untransformed and transformed individuals were infested with newly emerged larvae of the lepidopteran tobacco budworm, *Heliothis virescens*, a natural pest of tobacco. While the untransformed plants were almost completely devoured, the larvae on the transformed plants either died or failed to develop normally, producing only limited damage to the plants.

A second example of introduced resistance has exploited the properties of a toxin produced by the bacterium *Bacillus thuringiensis*, the so-called *B.t.* toxin. The toxin is a crystalline protein formed as a δ-protoxin during bacterial sporulation which, on ingestion by insects, is proteolytically cleaved to the mature toxin peptide. It is specifically toxic to lepidopteran insects and has been used in spray form as a crop protective agent. The toxin produced symptoms, in susceptible insects, of paralysis of the mouth parts, followed by a reduction or inhibition of feeding, and eventually death. Several genes encoding *B.t.* toxins have been isolated and characterized, and have recently been used to genetically transform plants, with the aim of conferring resistance. Fischoff *et al.* (1987) have introduced truncated *B.t.* toxin genes into tomato plants, and expression of the introduced genes resulted in toxicity to larvae of tobacco hornworm (*Mandura sexta*), *Heliothis virescens* and *Heliothis zea*. Different strains of *B. thuringiensis* produce toxins with different specificities, and their screening and/or structural modification by genetic manipulation may increase the range of activities.

A slightly different approach is the transfer of the *B.t.* toxin gene to soil bacteria which form associations with the roots of crop plants. The rationale here is that sprayed toxins are rapidly degraded, and continuous production by root colonizing microorganisms might protect the plant. The advantage of this system is the ease of transforming bacteria compared with plants. Although *B.t.* toxin genes have been transferred to *Pseudomonas fluorescens* strains able to colonize maize roots, permission for field trials has not yet been granted (Davison, 1988).

A number of other possible genes exist to confer resistance to insects. These can be classed as encoding either insecticidal toxins (cf. *B.t.* toxin) or seriochemicals, which are molecules which modify pest behaviour. Alternatives to *B.t.* toxin include insect neuropeptides, such as proctolin, which are active at very low (fmol) levels, and proteinaceous spider venoms. There are three principal groups of seriochemicals worthy of mention: pheromones, anti-feedants and host-recognition compounds.

Pheromones. These are volatile chemicals which are synthesized by insects as communication signals between individuals, and influence group activities such as feeding and mating behaviour. If plants could be engineered to synthesize specific pheromones, insect foraging and reproduction, and subsequent larval development on the plants, may be inhibited.

Anti-feedant chemicals. These, for example the terpenoid farnesene, are produced by some aphid-resistant plants such as *Solanum berthaultii* (in glandular leaf hairs), but not by the related but susceptible potato (*S. tuberosum*). These compounds act by eliciting an 'attack response' in aphids, causing them to scatter, and so prevent their establishment on the plant.

Host recognition. Finally, nematodes, the cause of numerous diseases of crops, interact very specifically with their host plant, and the interactions may be mediated by recognition molecules such as lectins (glycoproteins) on the plant cell surface, or by specific substances released into the soil. It may be possible in the future either to modify the synthesis or recognition properties of these chemical signals to inhibit nematode attack.

Resistance to Herbicides

The use of herbicides to control weeds has become an important part of agricultural practice. Efforts are in general directed towards the production of herbicides which are selectively toxic to weed species and (although not always successfully) to environmentally safer chemicals. Some success has been achieved concerning selectivity; for example, the herbicides 'Glean' and 'Oust', made by Dupont, control weed development at concentrations of 10–20 g/ha, while cereal crops exhibit considerable tolerance. However, this selective insensitivity is restricted to only a few species, and a number of herbicides are equally toxic to both crop and weed (e.g. 'Roundup' produced by Monsanto). Nevertheless, stably-resistant weeds do occasionally arise in the field, and resistant cultured cell lines can be obtained by selection *in vitro*. This has led to an investigation of the mode of action of herbicides and mechanisms of resistance, which potentially could be introduced into sensitive plants by genetic engineering techniques.

Four main groups of herbicides have been studied in this respect: (1) the triazine type; (2) the sulphonylurea and the imidazolanone types; (3) glyphosate; and (4) phosphinothricin. Strictly speaking, any chemical which can kill a plant can be considered to be a herbicide, including antibiotics which are used as selective agents in genetic transformation studies (Comai and Stalker, 1986). Conventionally, however, these are considered separately, and were discussed in relation to selection systems in Chapter 6. For a given phytotoxic compound, resistance to the effects could be mediated through one or more of a number of mechanisms: the plant or plant cell may either not take up the molecule, or may sequester it in a subcellular compartment, so preventing contact with a sensitive target (a protein, organelle etc.); the compound may be detoxified enzymatically, either by degradation or by combination with a second molecule, to produce a non-toxic product or products; resistant cells might produce a modified target (e.g. an enzyme) which is itself insensitive to the herbicide; and, finally, the normal target might be produced in abnormally large amounts, so that the herbicide concentration becomes a limiting factor and the plant (or plant cell) survives. The different modes of action and resistance mechanisms may be illustrated by some specific examples.

TRIAZINE HERBICIDES

Resistance to the triazine herbicides atrazine and simazine has been found in weeds growing close to areas of crop spraying, and studies of tolerant individuals and their progeny have demonstrated the resistance trait to be inherited maternally. Further analysis has found that protection is determined specifically by the *psbA* gene of the chloroplast, which encodes a 32-kDa protein of the photosystem-II electron transport system. It is this protein to which the herbicide binds, resulting, in susceptible plants, in the blocking of quinone/plastoquinone oxidoreductase activity, essential for driving electron transport. Resistant weeds have a much reduced, typically a 1000-fold lower, binding capacity for atrazine, and sequencing of the primary structure of the 32-kDa protein indicates that the differential binding capacity in susceptible plants is the consequence of relatively

Table 9.2 Atrazine resistance and mutation in the 32-kDa chloroplast protein*

Species	*Amino acid position*	*Mutation*	*Relative resistance to atrazine*	*Electron transport*
Amaranthus hybridus	264	Ser→Gly	1000 ×	altered
Solanum niger	264	Ser→Gly	1000 ×	altered
Chlamydomonas reinhardii (DCMU4)	264	Ser→Ala	100 ×	altered
C. reinhardii (Ar204)	?	?	84 ×	altered
C. reinhardii (Ar207)	255	Phe→Tyr	15 ×	normal
C. reinhardii (Dr2)	219	Val→Ile	2 ×	normal
C. reinhardii (BR202)	?	?	1 ×	normal

*From information in Comai and Stalker (1986) and Mets *et al.* (1985).

minor alterations in the amino acid composition. In a wide range of species the structures of the *psbA* gene and protein product are extremely highly conserved, and are identical in spinach and tobacco. However, atrazine-resistant *Amaranthus hybridus*, for example, differs from all sensitive species studied in that the usual serine at position 264 is replaced by a glycine residue; and identical and other mutations have also been discovered in resistant individuals of the other species (see Table 9.2). Genetic evidence suggests that these structural changes are due to single mutations in the *psbA* gene. Although in some cases the mutation does not adversely affect normal electron transport, in other cases (such as for the serine to glycine substitution) resistant plants may grow poorly, compared to their wild-type relatives, in a herbicide-free environment. Therefore, in attempts to engineer triazine resistance in susceptible cultivars, a balance must be drawn between yield gain through reduced competition from weeds and yield loss through defective electron transport.

Mets *et al.* (1985) have discussed a number of possibilities for genetically engineering resistance to triazines. In general, these can be classed as: (1) the transfer of an isolated mutant allele of the *psbA* gene; (2) the transfer of resistant chloroplast genomes, either by traditional plant breeding or by somatic hybridization; and (3) the induction of resistance by mutagenesis. In a strategy which involves the transfer of a mutant *psbA* gene, it is likely that the endogenous sensitive allele should be concomitantly inactivated or removed before full phenotypic expression of the introduced trait could occur. This is because the turnover of herbicide-bound proteins (predominantly the 'sensitive' ones) is reduced, and in a chloroplast containing both sensitive and resistant proteins, the herbicide would selectively stabilize the sensitive proteins, leading to a gradual replacement of the resistant protein, resulting in regression to a wild-type phenotype. Two approaches to this have been suggested: the sensitive gene could be replaced with the mutant gene by homologous recombination, solving the introduction of the new gene and removal of the wild type in a single step; the naturally-occurring recombination of mitochondrial DNA was discussed in Chapter 8 in relation to cytoplasmic sterility, and it is conceivable that similar mechanisms may operate in chloroplasts, and could be exploited. Alternatively genes could be introduced into the nucleus and the modified protein transported to the chloroplast; inactivation of the sensitive allele is then a separate problem, perhaps to be solved by mutagenesis. In both strategies the chloroplast envelope is a barrier, in the first to DNA uptake and in the second to protein transport. The latter could be solved by adding a sequence encoding a chloroplast-specific transit peptide to the resistance gene (see Chapter 3), and there is now some evidence that *Agrobacterium*-mediated transformation can produce non-homologous integration of DNA into the plastome. Chloroplast transformation has recently been aided by the development of a streptomycin resistance gene (streptomycin phosphototransferase) which can be used as a selectable marker. Streptomycin inhibits normal chloroplast replication, and unlike other antibiotics such as kanamycin, will not kill the cells in which only a proportion of the chloroplast population is transformed, so allowing the transformed organelles to replicate. Visual selection can be carried out, with transformed chloroplasts remaining green and untransformed ones bleaching in the presence of streptomycin (Jones *et al.*, 1987). This technique may go some way to solving the problem of promoting the proliferation of atrazine-resistant chloroplasts by co-transformation of cells with both streptomycin phosphotransferase and resistant *psbA* genes: although herbicide-binding sensitive organelles may be stable in the presence of atrazine, they will survive in the presence of the antibiotic, which, however, will select for the atrazine-resistant chloroplasts.

The alternative approach, recently reported by Cheung *et al.* (1988), appears to circumvent the problem of introducing mutant chloroplast genes into those organelles by instead 'converting' them into nuclear genes. The *psbA* gene from an atrazine-resistant biotype of *Amaranthus hybridus* was modified by linking its coding region to the transcription regulation- and transit-peptide-encoding sequences of the nuclear gene for the small subunit of Rubisco. The chimaeric gene was introduced into herbicide-sensitive tobacco plants by *Agrobacterium* transformation, with the result that some of the transgenic plants exhibited atrazine resistance. Moreover, the modified 32 kDa protein was demonstrated to be imported into the chloroplast and to function in photosynthesis. These experiments demonstrate, therefore, the possibility of modifying chloroplast-encoded genes by nuclear genome transformation.

SULPHONYLUREA AND IMIDAZOLANONE HERBICIDES

The Dupont herbicides 'Glean' and 'Oust' contain, as their active ingredients, the sulphonylurea-type compounds chlorsulphuron and sulphometuron methyl, respectively (Fig. 9.4). Both compounds have been shown to inhibit cell division both in root tips and in cultured cells, although it is possible to isolate resistant cell

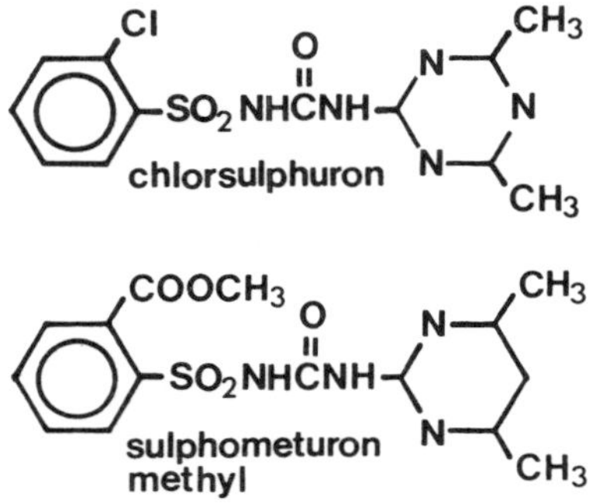

chlorsulphuron

sulphometuron methyl

Fig. 9.4 Structures of two sulphonylurea herbicides.

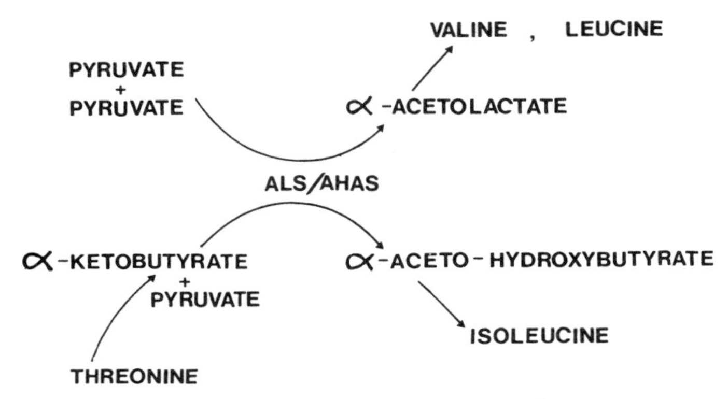

Fig. 9.5 Role of ALS/AHAS in amino acid metabolism.

lines by selection *in vitro*, from which resistant plants can be regenerated. Resistance was found, in some isolates, to be determined by the mutation of a single dominant or semi-dominant nuclear gene, suggesting that the herbicide target was a single protein. Studies with bacteria indicated that sensitivity to sulphonylureas could be alleviated by the inclusion of amino acids in the medium, indicating that the herbicides interfered with some aspect of amino acid metabolism. Specifically, the inhibition of growth of *Salmonella typhimurium*, *Saccharomyces cerevisiae* and also of peas (*Pisum sativum*) by sulphonylureas on minimal media could be prevented by supplying the branched-chain amino acids leucine, isoleucine and valine. These observations suggested that the herbicide target was the enzyme acetolactate synthase (ALS), which is also known as acetohydroxy acid synthase (AHAS). The reason for the two names relates to the dual function of the enzyme (Fig. 9.5), which catalyses both the condensation of two pyruvate molecules to α-acetolatate (which is further metabolized to valine and leucine) and the condensation of α-ketobutyrate and pyruvate to α-aceto-α-

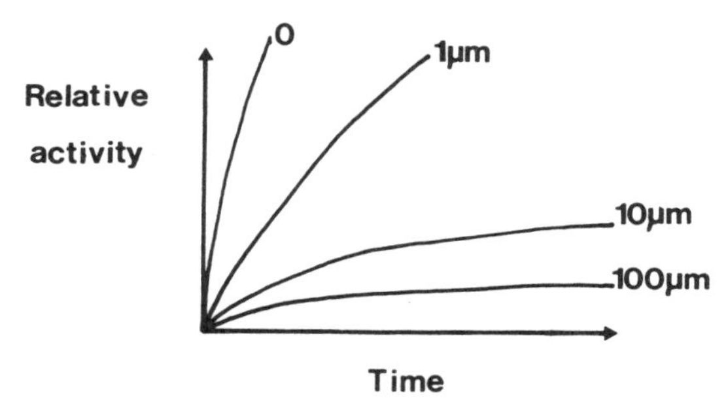

Fig. 9.6 Relative inhibition of bacterial ALS activity by increasing concentrations of sulphometuron methyl.

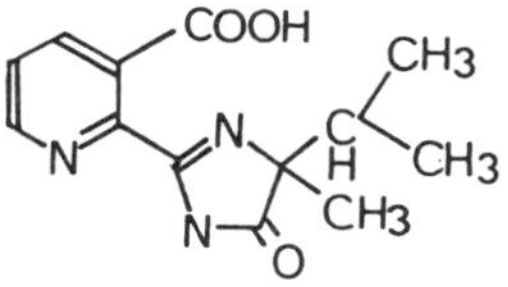

Fig. 9.7 Imazapyr, an imidazolanone.

hydroxybutyrate (from which isoleucine is derived). Assays of enzymes extracted from a range of both plants and organisms have confirmed the view that ALS is the target, the sulphonylureas acting as non-competitive inhibitors of the enzyme (Fig. 9.6). In assays of yeast enzyme, 5% inhibition of the enzyme occurs at a concentration of 120 nM sulphometuron methyl, but at only 20–30 nM for plants.

Resistance in bacteria can operate both by (1) over-production of ALS, due to the presence of the encoding genes *ILV2* or *ilvG* on high copy number plasmids, or (2) by the production of isoenzymes of ALS which are insensitive to the herbicide, encoded by the mutant *ILV2* genes. Insensitive forms of ALS have been identified in resistant mutants of tobacco, and the genes have been isolated using yeast ALS cDNA as a probe. Haughn *et al.* (1988) have found that the DNA sequence of a mutant gene from *Arabidopsis* differs from that of the wild type by a single base-pair substitution, and when transferred to chlorsulphuron-sensitive tobacco plants conferred resistance to the herbicide.

ALS is also the target enzyme of two other groups of herbicides, the imidazolanones and the triazopyrimidines or sulphonanilides, which are structurally unrelated to the sulphonylureas (Fig. 9.7). Presumably plants possessing a mutant enzyme conferring resistance to sulphonylureas would also reduce the inhibitory effects of these and perhaps other, as yet undiscovered, herbicides which exert their effect at the same metabolic step, but this is yet to be demonstrated. The fact that the sulphonylureas, imidazolanones and sulphonanilides bear no obvious similarities to the natural substrates of ALS raises many questions concerning

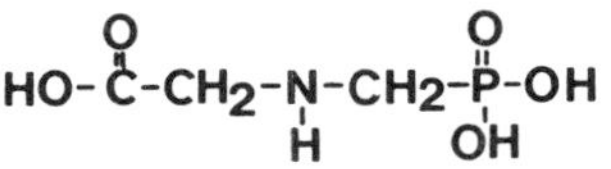

$$HO-\overset{\overset{O}{\|}}{C}-CH_2-\underset{\underset{H}{|}}{N}-CH_2-\overset{\overset{O}{\|}}{\underset{\underset{OH}{|}}{P}}-OH$$

Fig. 9.8 Structure of glyphosate.

their mode of action (which appears to involve kinetically complex binding) and concerning the evolution of the ALS binding site.

GLYPHOSATE

Glyphosate (*N*-phosphonomethylglycine, Fig. 9.8) is the active ingredient of 'Roundup', a wide spectrum herbicide produced by the Monsanto Company, which is toxic not only to plants but, like the sulphonylureas, also to fungi and bacteria. The observation that sensitivity to the herbicide can be overcome by the exogenous application of aromatic amino acids led to the proposition that glyphosate interferes with the synthesis of these compounds, resulting in cell death; and further biochemical studies have identified the target as the enzyme 5-enolpyruvylshikimate-3-phosphate synthase (EPSP synthase), also (but less commonly) known as 3-phosphoshikimate-1-carboxyvinyl transferase. This enzyme catalyses the condensation of phosphoenolpyruvate and shikimate-3-phosphate to form 5-enolpyruvylshikimate-3-phosphate, a precursor of the amino acids tryptophan, phenylalanine and tyrosine, and a range of aromatic secondary metabolites (Fig. 9.9). Glyphosate inhibits EPSP synthase activity by preventing phosphoenolpyruvate binding to the enzyme.

Bacterial EPSP synthase is encoded by the *aroA* gene, and mutant genes have been shown to induce resistance to glyphosate in, for example, *Salmonella typhimurium*, both by the over-production of the enzyme or by the production of an enzyme which is insensitive to the herbicide. Both resistance mechanisms have been exploited to generate herbicide tolerant plants. For example, Comai *et al.* (1985) have isolated a mutant *aroA* gene from *S. typhimurium*, which encodes an enzyme with a single proline to serine substitution, resulting in a decreased affinity for glyphosate. This gene was transferred to tobacco plants under the control of either the octopine synthase or mannopine synthase gene promoters, and expression of the chimaeric genes was correlated with increased resistance to sprayed glyphosate. EPSP synthase is encoded by a chloroplastic gene, but in this example resistance was obtained when the mutant gene was expressed in the cytoplasm. By fusing the transit peptide-encoding sequence of the Rubisco small subunit gene, however, the gene can be directed into the chloroplast, resulting in a 1000-fold higher level of resistance (Della Cioppa *et al.*, 1986).

A different approach was adopted by Shah *et al.* (1986). Here, the wild-type EPSP synthase gene coding region was linked to a constitutive promoter, i.e. that of the CaMV 35S RNA gene, and when transferred to *Petunia* plants, resulted in a 20–40 fold increase in activity of the enzyme, corresponding to a 10-fold increase in resistance to glyphosate. Presumably the targeting of the gene construct to the chloroplast would have resulted in an even higher tolerance.

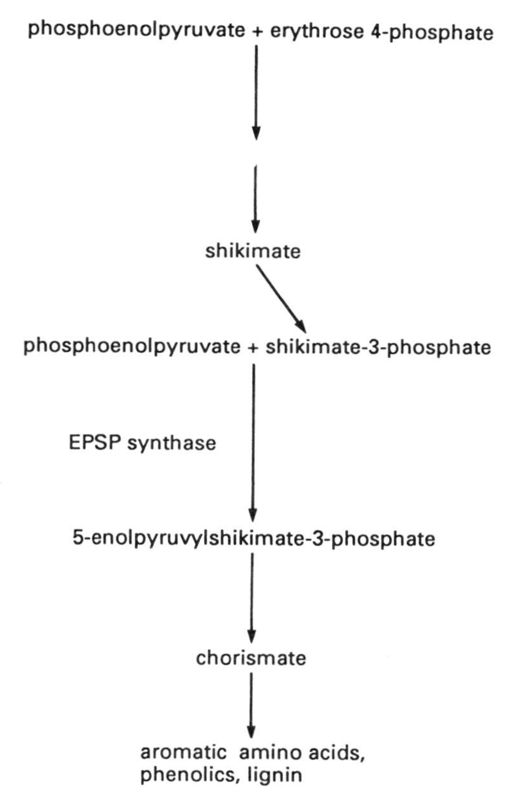

Fig. 9.9 Role of EPSP synthase in aromatic amino acid biosynthesis.

PHOSPHINOTHRICIN

L-Phosphinothricin (PPT) (Fig. 9.10) and bialaphos are relatively new herbicides, originally discovered as antibiotics produced by *Streptomyces* species. Bialaphos, produced by *S. hygroscopicus*, is a tripeptide comprising PPT, an analogue of L-glutamic acid and two residues of L-alanine. PPT appears to exert its effect as a competitive inhibitor of glutamine synthetase, which catalyses the conversion of glutamate to glutamine. This enzyme plays a key role in nitrogen metabolism in plants by virtue of the fact that it is the only enzyme capable of detoxifying ammonia generated by, for example, amino acid degradation or nitrate reduction. Inhibition of glutamine synthetase by the herbicide results in the accumulation of lethal levels of ammonia.

Bialaphos-synthesizing cells of *S. hygroscopicus* are resistant to its effects by producing a phosphinothricin acetyltransferase (PAT) which acetylates the free amino acid group of PPT. The *bar* gene which encoded PAT has been transferred to PPT-sensitive plants of tobacco, tomato and potato under the transcriptional

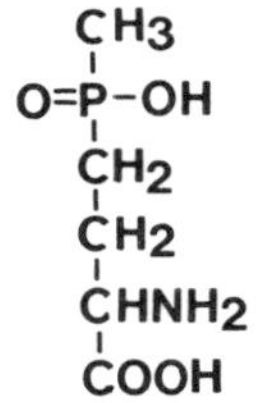

$$
\begin{array}{c}
CH_3 \\
| \\
O{=}P{-}OH \\
| \\
CH_2 \\
| \\
CH_2 \\
| \\
CHNH_2 \\
| \\
COOH
\end{array}
$$

Fig. 9.10 Structure of L-phosphinothricin.

control of the CaMV 35S promoter, and expression of the gene resulted in resistance to high levels of the herbicide (De Block *et al.*, 1987). The mechanism of resistance here is different from that of the other examples discussed, in that the herbicide is specifically detoxified by the introduced enzyme.

We have seen, then, a variety of approaches to engineer resistance to herbicides in plants: the generation of enzymes with altered sensitivity to the herbicide; the over-production of the wild-type enzyme; and the introduction of novel enzymes which detoxify the herbicide. Screening of plants and bacteria is likely to reveal other enzymes which may be useful in either catabolizing or conjugating herbicides so as to reduce their toxic effects, and some potential candidates are discussed by Comai and Stalker (1986).

Because of the possible environmental effects of the long-term accumulation of herbicides in the soil, it would be judicial for chemical companies to develop compounds which are readily degradable by soil microorganisms. Hopefully, engineered resistance will in any case lead to reduced levels of herbicides used, by the application of one non-selective, as opposed to several differentially selective, compounds.

Resistance to Stress

Stress can loosely be defined, in the context in which we will briefly consider it, as the adverse effects of diverse environmental factors on plant growth and/or development. In particular, we can discuss some of the responses to heat, cold, drought, flooding and UV irradiation, but the development of strategies to protect plants from such stresses by genetic engineering is currently rather difficult, and possible to contemplate only in general terms. If a limitation of nitrogen or other nutrients is considered to be an environmental stress, then it is relatively easy to see how genetic engineering techniques might be employed to alleviate the problem, for example, by manipulating nitrogen fixation (Chapter 7). However, the genetic, biochemical and physiological basis of resistance to other stresses, such as those imposed by heavy metal-containing soils or acid rain, are currently not well enough understood to consider genetic manipulation. The first step is to study the reaction of plants to stresses, and in some cases this is relatively well-characterized at the biochemical level.

FLOODING TOLERANCE

The primary stress involved in the flooding of non-aquatic plants is that of low oxygen availability. During anoxia, in which there is a complete lack of oxygen, there is a dramatic change in metabolism, with the accumulation of ethanol, alanine and glycolytic intermediates. There is a reduced range of proteins synthesized in anoxic tissues, but a characteristic increase in the activities of enzymes involved in fermentative glycolysis, including alcohol dehydrogenase, pyruvate decarboxylase, fructose-1,6-bisphosphate aldolase and glucose phosphate isomerase, and a transient increase in lactic acid production. Survival is dependent upon alcohol dehydrogenase activity in hypoxic tissues (in which there is some oxygen availability), so that partial fermentation occurs, but, while alcohol dehydrogenase activity increases, other survival strategies are involved, notably in increased lateral root formation and, in some species, aerenchyma production.

The regulation of alcohol dehydrogenase (ADH) is under transcriptional control, the synthesis of the mRNA being induced by low oxygen tension. Howard *et al.* (1987) have linked the 5′ sequences of the maize *ADH-1* gene to the coding region of the CAT marker gene and demonstrated its inducibility by anaerobiosis in a tobacco protoplast transient assay system. While these studies give insight into the regulation of inducible gene expression, the problem of how to engineer flooding tolerance remains, since the most important difference between tolerant and sensitive species is probably more related to their morphology (e.g. the presence of aerenchyma, lateral roots) than to their metabolic response.

HIGH AND LOW TEMPERATURES

The 'cardinal' temperatures, i.e. the maximum and minimum temperatures beyond which a particular metabolic function is undetectable, will vary widely between species, according to their ecological adaptation. Some aspects of the adaptations to extremes of temperature are genetically determined, but the unravelling of these determinants for transfer to crop species is only at its most preliminary stages.

The deleterious effects of both high and low temperature on plants may be aggravated by an accompanying unavailability of water, either because of evaporation or freezing, or because of high soil salinity (see below). At high temperatures, the transpiration stream may be reduced through drought, and the cooling effect at the leaf surface will be correspondingly lost, so compounding the stress. Some species, such as members of the Cactaceae and Euphorbiaceae, are structurally and biochemically adapted to desert conditions, being able to maintain a high water content; and growth of *Opuntia* has been recorded at tissue temperatures of 56.5°C. Other species, such as the creosote bush (*Larrea divaricata*) can tolerate considerable dehydration and are able to recover from desiccation equivalent to a water loss of 50% of their dry weight. A list of some morphological adaptations to high temperature is presented in Table 9.3.

Drought inflicts crop losses on approximately 40–60% of the world's cultivated

Table 9.3 Some morphological adaptations to high temperatures

Adaptation	
• Small, thin leaves	(prevents leaf temperature rising significantly higher than air temperature)
• High rate of transpiration	
• Pale colour of aerial parts	(reflect sun's rays)
• Hairs/waxes on aerial parts	
• Orientation of leaves to avoid a maximum absorption of sun's rays	
• Insulatory layer of corky bark	
• High osmotic potential of cytoplasm	

land. Since the biochemical determinants of drought tolerance are unclear, strategies of *in vitro* selection have been developed in a number of laboratories. Cultivated cells are commonly selected against high molecular weight polyethylene glycol, an osmoticum which is used to mimic natural drought stress, and the aim is to regenerate plants which are resistant to its detrimental effects on cell growth. A second strategy is to screen somaclonal variants, produced, for example, as somatic hybrids between a crop and a wild relative, or after rescue of sterile zygotic embryos (see Chapter 4), for improved drought resistance. This approach has been adopted for a number of important crops of arid climates, such as wheat, pigeonpea (*Cajanus cajan*, a high protein legume) and even rice.

Physiological studies have demonstrated that plant tissues subjected to water stress by drought, osmotic- or salt-stress, and also desiccating seeds (Chapter 8) characteristically increase endogenous levels of abscisic acid (ABA). This growth regulator in turn appears to induce the synthesis of specific proteins in these tissues, which may play a protective role. The precise role of ABA action is not yet understood, but a gene for an ABA-induced protein has recently been characterized (Marcotte *et al.*, 1988), and regulatory *cis*-acting sequences, inducible by exogenous ABA have been identified. This is an important first step in our understanding of the molecular basis of water-stress tolerance.

Apart from desiccation, high and low temperatures can cause injury through changes in the fluidity of membranes, disrupting ion transport, denaturation (by heat) or precipitation (by cold) of proteins, and by inducing imbalances in normal metabolism, with some pathways being more sensitive to temperature stress than others. Ice crystal formation can lead to intracellular dehydration and mechanical damage. There is also an accumulating body of evidence that extremes of temperature can elicit a change in the pattern of gene expression, with certain 'housekeeping' genes being switched off and other genes being transcribed in response to stress conditions. For example, a number of temperate plant species have been shown to synthesize up to about 20 proteins in response to incubation at temperatures of between 39–41°C. The appearance of these 'heat-shock proteins' (hsp) is very rapid, some accumulating to detectable levels within about 20 min, and their synthesis is *de novo* (e.g. Schoffl and Key, 1983). The upstream sequences of soybean hsp genes bear some homology to those found in *Drosophila*, which have been shown to regulate, in a heat-inducible way, the expression of marker genes in transgenic plants.

The physiological roles of hsp are not well understood. Their synthesis and accumulation is transient and can be induced by other stress conditions such as anaerobiosis, salt, osmotic stress, and heavy metals. There is, nevertheless, evidence that they are not simply a *symptom* of heat stress but play a protective role: a pretreatment of plants which leads to the accumulation of the proteins allows subsequent survival at normally lethal temperatures. Specific proteins are also correlated with the acquisition of freezing tolerance, but again no special function has been assigned to them.

One attempt to control frost damage to crops has been surrounded by controversy, because it relies on the release of a genetically engineered micro-organism. The bacterium *Pseudomonas syringae* causes frost damage by synthesizing a protein which acts as a nucleus for the formation of ice crystals at temperatures which are normally unfavourable for ice formation. A strain of *P. syringae* has been engineered in which the gene for the ice nucleation protein is deleted ('ice-minus' strain), and it is reasoned that, if this strain could compete with the wild type to predominate in the environment, frost damage could be reduced. Preliminary results have indicated that potato plants sprayed with the ice-minus strain showed 80% less damage than control plants during two early season frosts.

SALT STRESS

The molecular basis of salt tolerance, a complex polygenic trait, is not well understood. Efforts to understand the regulation of the response to high salinity or low pH (due to the pressure of high concentrations of Al^{3+} ions) have made use of cell culture systems with the development of salt resistant lines, and of studies at the whole plant level (Ramagopal, 1987). The approach has been to identify differences in gene expression in differentially sensitive cell lines of genotypes following salt treatment, and while specific polypeptides can be identified which are synthesized in relatively tolerant cells or tissues, no relevant genes have yet been isolated. The situation is complicated by the apparent effect of tissue organization in determining to some extent the resistant phenotype. Members of the halophytic Chenopodiaceae, such as *Salicornia*, *Beta* and *Chenopodium* commonly accumulate betaines, which appear to be stress-related compounds, and the study of their synthesis may be fruitful. An understanding of the molecular basis of the role of membrane-bound ion transport proteins in regulating the osmotic balance of the cell would also seem to be relevant.

ULTRA-VIOLET RADIATION

Radiation between about 260–400 nm is described as ultra-violet (UV), and is capable of inducing genetic mutations. Its significance as an ecological factor in reducing plant survival is probably not great, because structural and biochemical mechanisms appear to have been evolved to combat its effects. These include hairs and waxes on the epidermis to scatter the incident light, giving many desert species, for example, a blue sheen and thick cuticles which are impermeable to UV. A UV-blue photoreceptor has been described (see Chapter 8), and *in vitro* studies have demonstrated that enzymes involved in phenolics metabolism, notably PAL, 4-CL, CHS and CHI, are elicited by UV light (see above). Flavonoids and furanocoumarins absorb UV light, and presumably play a protective role.

General Reading

Callow, J.A., Estrada-Garcia, M.T. and Green, J.R. (1987). 'Recognition of non-self: the causation of disease', *Ann. Bot.* **60**, Suppl. **4**, pp. 3–14.

Collinge, D.B. and Slusarenko, A.J. (1987). 'Plant gene expression in response to pathogens', *Plant Mol. Biol.* **9**, pp. 389–410.

Comai, L. and Stalker, D. (1986). 'Mechanism of action of herbicides and their molecular manipulation', *Oxf. Surv. Plant Mol. Cell Biol.* **3**, pp. 166–195.

Davison, J. (1988). *'Plant beneficial bacteria'*, *Bio/Technol.* **6**, pp. 282–286.

Dixon, R.A. (1986). 'The phytoalexin response: elicitation, signalling and control of host gene expression', *Biol. Rev.* **61**, pp. 239–291.

Dixon, R.A., Bailey, J.A., Bell, J.N., Bolwell, G.P., Cramer, C.L., Edwards, K., Hamdan, M.A.M.S., Lamb, C.J., Robbins, M.P., Ryder, T.B. and Schuch, W. (1986). 'Rapid changes in gene expression in response to microbial elicitation', *Phil. Trans. R. Soc. Lond.* **B314**, pp. 411–426.

Marcotte, W.R., Bayley, C.C. and Quatrano, R.S. (1988). 'Regulation of a wheat promoter by abscisic acid in rice protoplasts', *Nature* **335**, 454–457.

Mets, L., Galloway, R.E. and Erickson, J.M. (1985). 'Prospects for genetic modification of plants for resistance to triazine herbicides', in *Biotechnology in Science*, Eds Zaitlin, M., Day, P. and Hollaender, A., pp. 301–312. Orlando, Florida, Academic Press.

Zaitlin, M. and Hull, R. (1987). 'Plant virus–host interactions', *Ann. Rev. Plant Physiol.* **38**, pp. 291–315.

Specific Reading

Abel, P.P., Nelson, R.S., De, B., Hoffman, N., Rogers, S.G., Fraley, R.T. and Beachy, R.N. (1986). 'Delay of disease resistance in transgenic plants that express the tobacco mosaic virus coat protein gene', *Science* **232**, pp. 738–743.

Baughman, G.A., Jacobs, J.D., and Howell, S.H. (1988). 'Cauliflower mosaic virus gene VI produces a symptomatic phenotype in transgenic tobacco plants', *Proc. Natl. Acad. Sci. USA*, **85**, pp. 733–737.

Cheung, A.Y., Bogorad, L., van Montagu, M. and Schell, J. (1988). 'Relocating a gene for herbicide tolerance: a chloroplast gene is converted into a nuclear gene', *Proc. Natl. Acad. Sci. USA* **85**, pp. 391–395.

Collinge, D.B., Mulligan, D.E., Dow, J.M., Scofield, G. and Daniels, M.J. (1987). 'Gene expression in *Brassica campestris* showing a hypersensitive response to the incompatible pathogen *Xanthomonas campestris* p.v. *vitians*', *Plant Mol. Biol.* **8**, pp. 404–411.

Comai, L., Facciotti, D., Hiatt, W.R., Thompson, G., Rose, R.E. and Stalker, D.M. (1985). 'Expression in plants of a mutant *aroA* gene from *Salmonella typhimurium* confers tolerance to glyphosate', *Nature* **317**, pp. 714–744.

De Block, M., Botterman, J., Vandewiele, M., Dockx, J., Thoen, C., Gossele, V., Rao Movva, N., Thompson, C., van Montagu, M. and Leemans, J. (1987). 'Engineering herbicide resistance in plants by expression of a detoxifying enzyme', *EMBO J.* **6**, pp. 2513–2518.

Della Cioppa, G., Bauer, S.C., Klein, B.K., Shah, D.M., Fraley, R.T. and Kishore, G.M. (1986). 'Translocation of the precursor of 5-enolpyruvyl-shikimate-3-phosphate synthase into chloroplasts of higher plants *in vitro*', *Proc. Natl. Acad. Sci. USA* **83**, pp. 6873–6877.

Fischoff, D.A., Bowdish, K.S., Perlak, F.J., Marrone, P.G., McCormick, S.M., Niedermeyer, J.G., Dean, D.A., Kusano-Kretzmer, K., Mayer, E.J., Rochester, D.E., Rogers, S.G. and Fraley, R.T. (1987). 'Insect tolerant transgenic tomato plants', *Bio/Technol.* **5**, pp. 807–813.

Gerlach, W.L., Llewellyn, D. and Haseloff, J. (1987). 'Construction of a plant disease resistance gene from the satellite RNA of tobacco ringspot virus', *Nature* **328**, pp. 802–805.

Harrison, B.D., Mayo, M.A. and Baulcombe, D.C. (1987). 'Virus resistance in transgenic plants that express cucumber mosaic virus satellite RNA', *Nature* **328**, pp. 799–802.

Haughn, G.W., Smith, J., Mazur, B. and Somerville, C.R. (1988). 'Transformation with a mutant *Arabidopsis* acetolactate synthase gene renders tobacco resistant to sulfonylurea herbicides', *Mol. Gen. Genet.* **211**, pp. 266–271.

Hilder, V.A., Gatehouse, A.M.R., Sheerman, S.E., Barker, R.F. and Boulter, D. (1987). 'A novel mechanism of insect resistance engineered into tobacco', *Nature* **330**, pp. 160–163.

Howard, E.A., Walker, J.C., Dennis, E.S. and Peacock, W.J. (1987). 'Regulated expression of an alcohol dehydrogenase 1 chimeric gene introduced into maize protoplasts', *Planta* **170**, pp. 535–540.

Jones, J.D.G., Svab, Z., Harper, E.C., Hurwitz, C.D. and Maliga, P. (1987). 'A dominant nuclear streptomycin resistance marker for plant cell transformation', *Mol. Gen. Genet.* **210**, pp. 86–91.

Kearney, B., Ronald, P.C., Dahlbeck, D., Staskawicz, B.J. (1988). 'Molecular basis for evasion of plant host defence in bacterial spot disease of pepper', *Nature* **332**, pp. 541–543.

Loesch-Fries, L.S., Merlo, D., Zinnen, T., Burhop, L., Hill, K., Krahn, K., Jarvia, N., Nelson, S. and Halk, E. (1988). 'Expression of alfalfa mosaic virus RNA 4 in transgenic plants confers virus resistance', *EMBO J.* **6**, pp. 1845–1851.

Lutgenberg, B. (Ed.) (1986). *Recognition in Microbe–Plant Symbiotic and Pathogenic Interactions*, Berlin, Springer-Verlag.

Mehdy, M.C. and Lamb, C.J. (1987). 'Chalcone isomerase cDNA cloning and mRNA induction by fungal elicitor, wounding and infection', *EMBO J.* **6**, pp. 1527–1533.

Ramagopal, S. (1987). 'Molecular biology of salinity stress: preliminary studies and perspectives', in *Tailoring Genes for Crop Improvement: an Agricultural Perspective*, Eds Breuning, G., Harada, J., Kosuge, T. and Hollaender, A., pp. 111–119. London, New York, Plenum Press.

Sanchez-Serrano, J., Keil, M., O'Connor, A., Schell, J. and Willmitzer, L. (1987). 'Wound-induced expression of a potato proteinase inhibitor II gene in transgenic tobacco plants', *EMBO J.* **6**, pp. 303–306.

Schoffl, F. and Key, J.L. (1983). 'Identification of a multigene family for small heat shock proteins in soybean and physical characterization of one individual gene coding region', *Plant Mol. Biol.* **2**, pp. 269–278.

Shah, D.M., Horsch, R.B., Klee, H.J., Kishore, G.M., Winter, J.A., Turner, N.E., Hironaka, C.M., Sanders, P.R., Gasser, C.S., Aykent, S., Siegel, N.R., Rogers, S.G. and Fraley, R.T. (1986). 'Engineering herbicide tolerance in transgenic plants', *Science* **233**, pp. 478–481.

Shahin, E.A. and Spivey, R. (1986). 'A single dominant gene for *Fusarium* wilt resistance in protoplast-derived tomato plants', *Theor. Appl. Genet.* **73**, pp. 164–169.

Thornberg, R.W., An, G., Cleveland, T.E., Johnson, R. and Ryan, C.A. (1987). 'Wound-inducible expression of a potato inhibitor II-chloramphenicol acetyltransferase gene fusion in transgenic tobacco plants', *Proc. Natl. Acad. Sci. USA* **84**, pp. 744–748.

Chapter 10

Prospects for the Future of Plant Biotechnology

Summary of the Current State of the Art

Plant biotechnology is a rapidly expanding area of biology, because it is the product of several disciplines which, until relatively recently, were considered separate: molecular biology, tissue culture, chemical engineering, plant pathology and, increasingly, accountancy. Ecological and political considerations will certainly feature heavily in future plans to release genetically-engineered organisms into the environment. Such diversity in subject matter only serves to make the writing of an admittedly, but inevitably, incomplete account of the field difficult. However, our aim has been to illustrate the interactions between the various disciplines, to indicate the types of experimental approach which have been adopted to solve specific problems in both fundamental and applied areas of plant biology, and to point to the directions in which we feel research will move.

In the last ten years or so, significant advances have been made which have laid out the basic strategies for genetically engineering a plant. During the early 1970s, regeneration of plants from protoplasts or single cells was a routine procedure for only a very small number of species, such as tobacco, carrot and *Petunia*, which have subsequently been exploited as 'model' species for genetic transformation studies. Since then, improvements in tissue culture techniques (such as the use of nurse cultures agarose culture) and, more importantly, of a fuller appreciation of the factors which influence the growth, differentiation and development of cells in culture, have extended the range of species which can be regenerated from callus to include legumes, woody species and a number of monocotyledonous species, notably the cereals (including rice, maize and wheat), which had previously been considered very difficult, if not impossible to manipulate *in vitro*. For the cereals, the turning point came with the observations that: (1) high levels of 2,4-D, the synthetic auxin, induce rapid callus proliferation at the expense of structural differentiation, but its removal may allow structural reorganization to occur; and (2) the source of explant material is crucial. Previous classical studies by Skoog and Miller had demonstrated clearly that the formation of roots or shoots could be induced from undifferentiated pith callus of tobacco simply by altering the ratio of auxins to cytokinins in the medium. For many years this approach formed the basis of attempts to regenerate plants from the callus of monocotyledons, but with unsuccessful results. Callus produced from mature, fully-differentiated tissues, such as leaves, appeared to be 'inflexible' in its subsequent tissue culture response. The solution was discovered through the use of immature, undifferentiated and highly meristematic tissues, such as young embryos or inflorescences, as explant material. These tissues retain a morphogenetic capacity, quite undefined in molecular terms, and which, under appropriate culture conditions, give rise to an embryogenic callus. Morphologically, at least, this callus is distinguishable from non-embryogenic callus, being composed characteristically of densely-cytoplasmic and relatively slowly-dividing cells, and will give rise to somatic embryos, as opposed to the discrete roots or shoots commonly observed in the regeneration of dicotyledonous species. These somatic embryos can be grown up to plants.

More recently it has been shown to be possible to regenerate monocotyledon crop plants (maize, rice, wheat and sugarcane) from protoplasts, and it is expected that within the next two to five years the list will include other cereals, such as barley and sorghum. Success, essentially, has been due to the realization that the developmental flexibility of cereal cells is not retained for long after explantation and they rapidly enter a fully determined state in which their path of development cannot easily be altered. There appear to be 'time windows' in the development of the plant tissues during which period manipulation *in vitro* is relatively easy: the explanted cells can respond favourably to external (e.g. hormonal) influences. Outside these time windows, the cells will not respond. Therefore, in their successful attempt to regenerate maize plants from protoplasts, Rhodes *et al.* (1988) used embryogenic cells as starting material, and maintained the protoplasts and protoplast-derived cells in such a way as not to lose that developmental potential, using nurse cultures and appropriate growth media. The difficulty, of course, is to recognize those cells which are morphogenetically competent, and a great deal of effort has been put into finding biochemical and molecular markers unique to embryonic cells. As was discussed in Chapter 8, many genes are switched on (and others switched off) during embryogenesis, but no gene has yet been isolated which can be said to have a determinative role in development, i.e. which, when expressed, activates the embryogenic process. It is likely that no such single gene exists, and embryogenesis is under polygenic control; however, it has been possible to transfer the trait for 'improved regeneration efficiency' from one tomato cultivar to another, indicating that the behaviour of cells in culture is influenced by the genetic composition of the explant tissue; and many examples exist in which the precise genotype of a species used had a pronounced effect on callus formation, protoplast plating efficiency, and frequency of transformation by *Agrobacterium*. If 'regeneration genes' could be identified and their regulatory effects defined, then some real meaning could be attached to the terms 'developmental competence' and 'determination'; and the study of the expression of such genes in transgenic plants, and in particular their effects on the frequency of regeneration and transformation, is an exciting, if long-term, prospect.

On a more practical level, tissue culture techniques have provided means of vegetatively producing large numbers of 'clones' by micropropagation, of generating virus-free material by meristem culture, and of regenerating hybrid plants from embryos which would otherwise abort (embryo rescue). The large-scale culture of plant cells for the production of useful secondary metabolites has produced one commercial success, the Mitsui shikonin process, but in general it is fair to say that mass culture has failed to deliver results (of a commercial kind) in proportion to the effort put in. Immobilized cell and hairy root culture systems certainly hold much promise, because they both exploit the dependence of production on structural organization seen for very many secondary compounds. The molecular basis for this rationale is gradually emerging.

Techniques for the genetic transformation of plants have also advanced rapidly in the last few years, with the development of two broad strategies: *Agrobacterium*-mediated gene vectors, based on the Ti or Ri plasmids, and direct gene transfer which includes both chemically- and electrically-mediated introduction of DNA into protoplasts, and also injection, either into the nuclei of single cells or protoplasts, or into meristematic regions of the intact plant. The list of genetically-transformed plants now includes more than twenty species, although much fundamental research is carried out using model systems such as tobacco and *Petunia*. Progress with the cereals has not been rapid, and the prospects for improving the situation will be discussed below.

Our understanding of the mechanisms involved in *Agrobacterium*-mediated transformation have aided rapid advances in its exploitation in genetic engineering. The identification and elimination from the Ti plasmid of many of the DNA sequences unnecessary for experimental gene transfer, such as the growth-regulator genes, and the recognition that the virulence genes can act in *trans* to the T-DNA to regulate its excision, has led to the construction of small (and so easily cloned and stable) binary vector systems, which contain multiple cloning sites and selectable marker genes for facilitated engineering. Progress is also being made in direct gene transfer systems, with improved understanding of, for example, the effects of electroporation on plasma-membrane permeability, the fate of DNA within protoplasts (both before and after integration), and of factors which influence both transient and stably integrated gene expression. Transient expression systems are powerful tools for the rapid functional analysis of novel gene constructions, and direct gene transfer is of potential value for the stable transformation of species for which the *Agrobacterium* system is not readily amenable, such as monocotyledons; success, however, is dependent on the high frequency regeneration of plants from protoplasts.

Transformation techniques have also made an enormous impact on our ability to study the relationship between gene structure and function. Throughout this book we have quoted examples in which upstream or downstream regions of particular genes have been identified as having a regulatory role by linking them to the coding region of a marker gene and studying the expression of the chimaeric construct in protoplasts or transgenic plants. The results have been particularly impressive if the regulatory sequences were derived from inducible, tissue-specific or developmentally specific genes, and such sequences are now themselves being used as tools to investigate the expression of other genes; and may also prove to have useful applications, for example, in switching on resistance genes during periods of insect or pathogen attack.

The applications of other molecular biological techniques, such as cDNA cloning, *in vitro* translation, and gene library screening to basic questions of plant biology has resulted in significant advances in our knowledge of developmental processes and of mechanisms of resistance to pests, diseases and herbicides. This knowledge has produced the most spectacular examples to date of the design and construction of plants with novel phenotypes, i.e. genetic engineering in its true sense: the production of plants with introduced resistance to herbicides, viruses and insect pests.

Other techniques are now emerging which offer exciting prospects for manipulating development and metabolism. These include the use of anti-sense RNA (Chapter 3) and ribozyme (Chapter 9), so-called 'hormone engineering', and transposable elements. Anti-sense RNA and ribozyme are potentially very powerful experimental tools because they could be used to inhibit the expression of particular genes in a highly specific way by binding to a specific mRNA. There are currently few published reports of the successful inhibition of gene expression by anti-sense RNA, and none as yet for ribozyme, although it is expected that more will appear over the next few years. The potential of both methods in conferring resistance to viruses has already been discussed, and applications could perhaps be found in protecting plants against fungi in blocking metabolic pathways which compete for precursors in the synthesis of a valuable secondary product, and in studying the role of specific genes in regulating differentiation and development, such as the homeotic genes of *Arabidopsis*. In general, therefore, anti-sense RNA and ribozyme could provide yet more tools for modifying metabolism and development.

An alternative method, which may also prove to be extremely useful in investigating and modifying developmental processes, is 'hormone engineering'. Here, genes which encode enzymes catalysing the synthesis of auxins or cytokinins, such as are found in wildtype *Agrobacterium* T-DNA, are transferred to plants in an attempt to modify development by altering the endogenous levels of the growth regulators. For example, Ooms and Lenton (1985) have introduced T-DNA cytokinin genes into potatoes and observed alterations in the number, sizes and maturity dates of tubers produced by the transgenic plants. Similarly, Klee *et al.* (1987) have introduced into *Petunia* and tobacco auxin biosynthetic genes under the control of either a constitutive or a tissue-specific promoter. When driven by the constitutive CaMV 19S promoter, one auxin gene, *iaaM*, resulted in a 10-fold over-production of indoleacetic acid (IAA) in transgenic plants, which correspondingly exhibited abnormal morphological features, such as strong apical dominance (little lateral branching), increased vascular tissue and modified leaf shape. Auxin receptor molecules and the encoding genes have now also been isolated, and their role in modifying the response of different cell types to endogenous growth regulators will no doubt be studied in transgenic plants with a view to engineering development.

A third novel approach to understanding gene regulation in development is by studying transposable elements (reviewed by Schwartz-Sommer and Saedler, 1987). There is evidence that these elements can modify the regulation of specific genes by insertion into sequences 5′ to the coding region, and a variety of effects can be achieved by a single family of elements; furthermore, for a given inserted sequence, its effect appears to depend on its site of insertion within the target gene. Thus, the insert can act as a *cis*-acting regulator and also as a receptor sequence for *trans*-acting DNA binding proteins. Perhaps transposable elements will one day be used to modify gene expression in a targeted way, or as vectors for other regulatory sequences, when more is understood about the way in which they are excised and integrated.

We have reached the point where, for a trait encoded by a single gene, it is relatively easy to isolate, clone and transfer that gene to a model species; and for traits of commercial importance, the availability of transgenic plants or even of a novel sequence of DNA can have a major impact on agricultural markets (as well as on the share value of plant biotechnology companies!). So far, so good, but as is always the case in biology, as soon as one limitation to the system is overcome, a second immediately takes its place. What are the current principal technical limitations to progress in plant biotechnology?

Current Limitations

We can identify at least three broad areas of basic research which need to be attacked before further major advances can be made; more specific problems, of a methodological or political nature, will be considered separately. The main problems relate to: (1) the genetic and molecular characterisation of traits which are determined by polygenes; (2) instability and variability of the response of tissues to manipulation *in vitro*; and (3) the general lack of detailed knowledge of plant metabolism and its regulation *in vivo*.

POLYGENIC TRAITS

Polygenes are responsible for the continuous variations observed for a given phenotype or developmental trait in a population of plants (or animals). The more genes that are involved in the expression of a trait, then the closer the pattern of variation corresponds to a normal distribution curve. In the simplest situation, each member of a set of polygenes can be considered to have an equal, but proportionally small, effect on the expression of the trait, and the effect of each

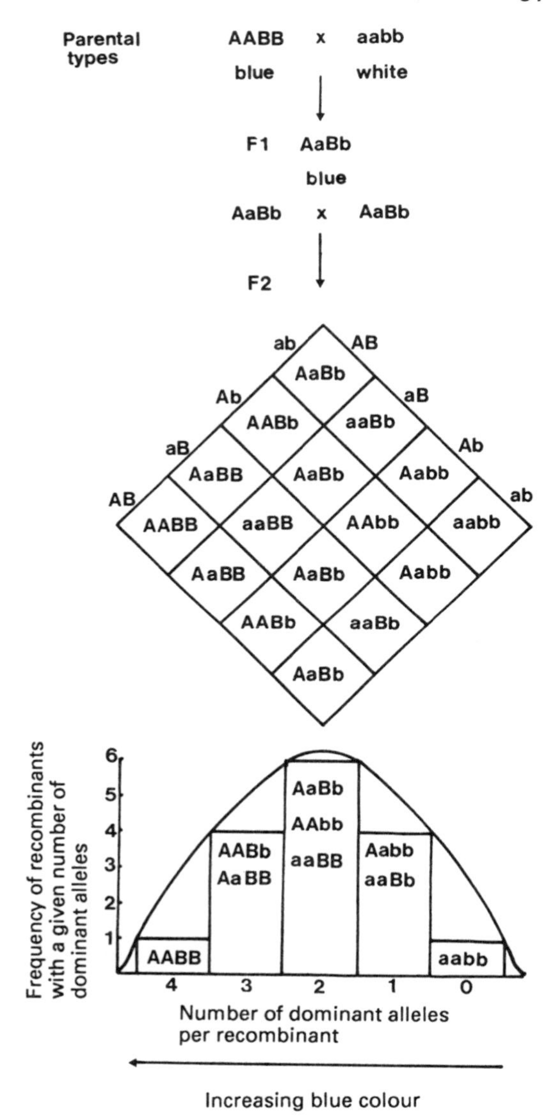

Fig. 10.1 A simple example of the effect of polygenes (in this case, two dominant genes, A and B, with recessive alleles a and b) on continuous variation of flower colour. The presence of one or more dominant alleles results in the production of some blue pigment in the flower, and the effect of increasing numbers of dominant alleles is additive. The variation in flower colour approximates to a normal distribution. With an increasing number of genes the parental recombinants become rarer.

gene is additive (see Fig. 10.1). Attempts to genetically manipulate a polygenic trait require the alteration of the structure or regulation of a large proportion, if not all, of the component genes, because the effect of altering or deleting a single gene is inversely proportional to the number of genes involved. For no polygenic trait has the number of component genes been determined precisely, and, with large numbers of these genes, the precise function of each is extremely difficult to determine. As indicated in Chapter 3, insertional mutagenesis is not likely to be a useful approach because the probability of simultaneously mutating a series of component genes without killing the cell is negligible. Perhaps the best hope is the use of restriction fragment length polymorphism (RFLP) techniques to correlate the inheritance of a trait with a particular fragment or set of fragments of DNA. If, for example, the genes for enzymes of phenylpropanoid metabolism or chitinases and $\beta(1\rightarrow3)$-glucanases, involved in aspects of disease resistance, were to be clustered, then perhaps they could be isolated and together transferred between plants. Certainly the genes for some microbial metabolic pathways such as antibiotic production or sulphur metabolism are grouped together, and analysis of the organization of genes for complex metabolic pathways in plants is greatly needed to determine whether such clustering occurs. In the meantime, the production of somatic hybrids by protoplast fusion, and partial genome transfer of irradiated protoplasts, is useful, if not particularly precise, for manipulating complex traits, and has been used to successfully introduce genes for resistance to potato leaf roll virus from *Solanum brevidens* into *S. tuberosum*. Although the hybrid plants failed to produce normal tubers, many were female fertile and could be introduced into a potato breeding programme.

INSTABILITY AND VARIABILITY

Instability and variability in the morphology, genetic complement and biosynthetic activity of cultured cells and tissues is now well documented and can be considered either beneficial or detrimental, according to the objectives of the work. The benefits of the variability within a population of cells include the possibility that variant or mutant cell lines, with specific useful properties, may be isolated by screening or selection. Useful properties might include a strong embryogenic potential, an ability to synthesize relatively high levels of a secondary metabolite, or resistance to a selective agent; and examples of all these have been discussed throughout this book. The observed heterogeneity in cell cultures is a function of the explant material (which comprises a range of tissue types) and also of the culture process itself, which ironically may largely account for the generation of instability *in vitro*. This is a problem in relation to the maintenance, through a series of subcultures, of the morphogenetic and biosynthetic potential of selected cell lines, and to the clonal micropropagation of valuable genotypes.

The control of instability requires detailed knowledge of its causes, as discussed in Chapter 5. Certain components of the medium, such as synthetic auxin 2, 4-D, are thought to have a damaging effect on karotype stability, either through directly mutagenic properties or by inducing rapid and disorganized cell divisions which may lead to aneuploidy. Meristematic cells with organized divisions appear to be less susceptible to genetic instability than friable cultures composed predominantly of vacuolated cells, and the reasons for this require investigation. It may be possible to select indirectly for genetically stable (or relatively stable) lines by selecting for small, densely-cytoplasmic cells. The fact that cereal somatic embryos generally are genetically uniform may reflect their regeneration from meristematic cells which are inherently stable (as opposed to the more friable, disorganized, faster-growing and non-embryogenic callus), or it may be simply that embryogenesis is capable of proceeding only from cells which happen to have a full genetic complement. More work, therefore, is needed to determine whether somaclonal variation is a function of the cell type from which the regenerant is derived, or whether embryogenic cells are as susceptible to mutation and chromosomal aberration as non-embryogenic cells.

If one accepts that a certain amount of instability is inevitable in long term cultures, the principal practical solutions are: (1) repeated screening or selection for the trait of interest, either from established cell lines or from freshly initiated cultures which may be genetically and biosynthetically less disrupted; and (2) the storage of material under non-growth conditions, while maintaining its viability, for example, by cryopreservation (freeze storage, Chapter 5). The techniques of cryopreservation are now well advanced and have been shown to be applicable to a wide range of species. As biotechnology generates valuable unique cell lines and genotypes the value of storage techniques will surely increase as a protection against accidental loss of cultures through contamination or power failure and to facilitate transfer between laboratories. The latter point raises the question of the accessibility of specific germplasm. There is an increasing requirement for an organized communication and distribution network to which laboratories (or companies) with useful cell lines or genotypes could become affiliated. Each laboratory would provide a detailed description of their available material, including its specific character(s) of interest, culture and storage requirements and so on, and the relevant information could be filed in a data bank, readily available to interested parties.

We have touched on the observation that the response of cells in culture may be influenced by the genotype of the plant which provides the explant material. While this has suggested that 'regeneration genes' might be isolated and transferred to recalcitrant species, it raises practical problems in that many genotypes may have to be screened before the desired response can be achieved. However, the role of genotype *per se* is not particularly clear, since the problems (e.g. lack of regeneration) in one genotype may be overcome by changing the culture medium or explant tissue, or by using younger cultures. Therefore, while the use of a particular genotype may be influential under a particular set of conditions, it must be recognized that it is only one of a number of factors contributing to the variability of response (see e.g. Vasil, 1988), and a greater understanding of all these variables is urgently required as time and money are invested in producing valuable cultures.

METABOLISM AND ITS REGULATION

The third major area which is limiting advances in plant biotechnology is the lack of available information on plant metabolism and its regulation. This becomes strikingly apparent when attempts are made to isolate genes involved in stress responses (except perhaps when the polypeptide gene product is itself active in protecting the plant, such as proteinase inhibitors or chitinases) or when attempting to manipulate the synthesis and accumulation of specific secondary metabolites in culture cells.

The problems are complex. Firstly, many of the metabolic intermediates of especially secondary metabolic pathways are unknown, and this in turn inhibits or prevents the study of regulatory aspects such as the identification and isolation of key enzymes. It may be possible to estimate the pool size(s) for a given intermediate by, for example, pulse chase studies, but other intermediates may be so rapidly metabolized that they may not accumulate to detectable levels. Inhibitor studies may be useful, but have to be interpreted with caution, as the full range of their effects may not be known. Furthermore, because metabolism is a network of interconnecting and to an extent interdependent pathways, the effect of manipulating one key enzyme may cause an imbalance in the flux of other pathways, which may be detrimental to the functioning of the cell. Finally, successful manipulation of a pathway depends on the level at which it is regulated *in vivo*. Although a given enzyme may be demonstrated to have particular characteristics when studied in a cell-free system, and thereby be implicated in having a regulatory role, this may not be a true reflection of its role *in vivo*. For example, although the synthesis of phenylalanine ammonia lyase (PAL) may be shown to be inducible by UV light or fungal elicitors, the 'uninduced' level of the enzyme in the cell may be adequate for the provision of the phenylpropanoid skeletons of many secondary metabolites, and its activity may be regulated *in vivo* by substrate supply. The genetic manipulation of the activity of an enzyme by, for example, altering the gene copy number or by manipulating the promoter region of the encoding gene requires that the activity of the enzyme is regulated by *de novo* synthesis at the transcriptional level and not, for example, by mRNA processing or stability, by post-translational covalent modification of polypeptide structure, or by substrate or cofactor availability to the mature enzyme. Similarly, manipulation of enzyme activity by modifying the polypeptide structure to increase its affinity for substrates or cofactors requires that they themselves are not limiting. These examples serve to illustrate the importance of understanding the activity and regulation of pathways *in vivo*, and undoubtedly the application of the newer techniques, such as nuclear magnetic resonance (NMR) spectroscopy of living tissues, which is non-invasive, will become increasingly important.

There is still a large hiatus in our knowledge relating to the role of *trans*-acting factors in the control of gene expression. We have summarized the current scarce evidence for the existence of sequence-specific DNA-binding proteins, the products of presumed 'regulatory genes', believed to act in the modification (either induction or suppression) of the transcription of specific 'regulated' genes, but the mystery remains – what controls the regulatory genes? There is evidence that growth regulators induce the transcription of specific genes, but it is generally believed that they do not interact directly with the genome (it is difficult to imagine, for example, how a single molecular structure such as that of IAA could differentially activate a range of genes), but rather through second messengers, such as the calcium/calmodulin system. The oligosaccharins, small carbohydrate fragments, are currently attracting interest as the 'new growth regulators', being able to elicit a range of physiological responses, including the activation of proteinase inhibitor genes, but their mode of action is also uncharacterized. Precisely how gene expression is linked biochemically to physical and chemical environmental signals is still one of the most fundamental problems of plant biology.

Specific Problems

Having outlined the general bottlenecks in progress in plant biotechnology, as we perceive them, we will now consider some of the more immediate hurdles.

Probably the most pressing problem is the development of reproducible, high-frequency transformation systems for the cereal crops. Some progress has been made, notably in the use of injection techniques to introduce naked DNA molecules into developing inflorescences of rye, resulting in the production of kanamycin-resistant seed. Information is still not available on the stability of the introduced genes over a series of generations, and there are no reports of the successful application of the method to other species. There is increasing hope for the use of *Agrobacterium* as a gene vector; the sensitive agro-infection assay system (Chapter 5) has clearly demonstrated that T-DNA can be transferred into cereals, and emphasis has been placed on the importance of introducing foreign genes into specific tissues, i.e. meristematic regions, to obtain maximum expression. There is also evidence that the natural host range of *Agrobacterium* strains can be extended by including cytokinin genes in the T-DNA, and extracts of wounded dicotyledonous tissues, which contain flavanoid compounds, such as acetosyringone (not found in monocotyledons), can induce the activity of the Ti-plasmid 'virulence' genes to allow T-DNA transfer to at least some monocotyledon species. Direct gene-transfer studies into maize protoplasts have indicated how intron sequences, inserted towards the 5′ end of the coding regions of both transiently expressed and stably integrated marker genes, can increase the levels of expression when under the control of both constitutive and inducible promoters (Callis *et al.*, 1987, see Chapter 3). Therefore, if these sequences, or other enhancers, or stronger promoters than the CaMV 35S, can be isolated and used to increase the activity of a selectable marker enzyme, such as neomycin phosphotransferase, higher levels of a selectable agent, such as kanamycin, could be used to detect transformants. This may be of particular value for the selection of putatively transformed cereals, which appear to have a high endogenous resistance to a range of antibiotics (Hauptmann *et al.*, 1988). Alternatively, more suitable selective agents could be found for use with cereals, such as some herbicides (e.g. phosphinothricin). Finally, the advances in cereal regeneration from protoplasts and somatic embryos has opened up the possibility of obtaining transformants by direct gene transfer. Transgenic plants of both maize and rice have now been produced from electroporated protoplasts, and for rice, are fertile (Shimamoto *et al.* 1989).

One of the limitations to current transformation techniques is that the integration of DNA is random. This results in a reduction in the efficiency of recovery of useful transformants, because the position in the genome at which integration occurs appears to influence strongly the level of expression of the introduced gene ('position effects'). Methods to target DNA sequences to particular nuclear, chloroplastic and mitochondrial genomes are therefore potentially particularly useful, but are not yet available. More work must be carried out to characterize fully the integration event. There is evidence for homologous recombination in eukaryotes, and the frequency of transformation may be increased by flanking genes with repetitive sequences. An alternative strategy of manipulating chloroplast- or mitochondrion-encoded gene products is to 'convert' the genes to nuclear sequences by exchanging the normal regulatory region for those of a nuclear gene, and target the gene product (rather than the gene itself) by the addition of transit peptide sequences.

It should be quite apparent by now that, while much headway has been made in our general understanding of gene structure and regulation, there are relatively few useful genes available. In this book we have discussed the main areas of current interest in this respect: resistance to diseases, pests and herbicides, product quality and yield and reproductive biology, but for the reasons we have considered above, few genetic determinants for these traits have been isolated. While considerable successes have been achieved in tackling resistance to viruses, insect pests and herbicides, other important objectives of genetic manipulation can be identified in which significant progress would have a great impact on world agriculture. These include: (1) the manipulation of cereal storage proteins to produce a grain with a balanced amino acid composition, so that a single crop can be used as a nutritious staple diet; (2) a fuller understanding of fungal pathogen-resistance mechanisms, and more information on the mode of viroid action, so that crop-protection strategies can be more effectively developed; (3) the engineering of nitrogen fixation in non-leguminous crops, so that lower levels of nitrate fertilizers, which may contribute to ecological problems, need be used; and (4) the expansion of the range of habitats in which a given crop species can successfully grow and develop. These are some of the more difficult problems, while for others, such as the manipulation of cytoplasmic male sterility or virus resistance, the way ahead is a little easier to see because of the large amount of information available.

In more general terms, a major strength of genetic engineering is the extension of genetic diversity in the crop plant gene pool (Jain, 1988). Genetic diversity is traditionally introduced into crops from wild relatives, and is essential, not only as a source of valuable traits to be exploited as human needs or tastes change, but also as a protection against epidemics, which can devastate in a genotype- or cytotype-specific way, or in increasing the adaptability of a crop to an alien environment. There is a danger that, as high-yielding genotypes are developed in Western agriculture, their use will become more widespread at the expense of others, resulting in a much reduced gene pool. For example, while Indian farmers are believed to have used, at one time, up to 3000 different varieties of rice, it is estimated that this number will have declined to a handful by the end of the century. Moreover, there is an emphasis in modern plant breeding on uniformity of product as a result of pressure from industry. Plant biotechnology can help this situation enormously, because it does not rely on Mendelian inheritance to recombine genes. The normal processes of sexual reproduction are bypassed, and genes from evolutionarily distant species can be introduced into a greeding programme.

Technology

Of the technological aspects, we can foresee advances in a number of areas. As more becomes known of the regulation of secondary metabolism, for example, the design of bioreactors will evolve to meet the requirements of cultured cells to maximize the production of specific natural compounds. While the only commercial operation, i.e. that of shikonin production by suspended cells of *Lithospermum erythrorhizon*, is a two-phase process (comprising a growth phase followed by a production phase, in two separate vessels), we envisage the prominent use of immobilized cell and hairy root culture bioreactors, in which a single reactor vessel is employed and the production phase is extended relative to the growth phase (see e.g. Webb and Mavituna, 1987). The advantages of these systems, for some species/products at least, are of both a physiological and a chemical engineering nature.

The mechanization of micropropagation is also likely to be developed in the next few years, being currently very labour-intensive and therefore expensive. Some progress in the use of robots has already been made, and will surely advance rapidly as the power of artificial intelligence systems increases.

More agricultural aspects of the technology include the prospective use of cultured cyanobacteria (blue-green algae), which can fix nitrogen, as 'bio-fertilizers'. These microorganisms can be grown up relatively cheaply in artificial ponds using sunlight as their energy source, and on harvesting can be dried to a dormant but still viable powder. This powder can be used to inoculate soil, where regrowth of the algae occurs, adding nitrogen to the soil. Cyanobacteria are also a source of β-carotene and other useful chemicals (Karuna-Karan, 1987).

The Impact of Plant Biotechnology on the Environment and Industry

By the end of the twentieth century, the world's population is expected to have increased from its present 5 billion to 6.2 billion, and it is estimated that world agriculture must increase its current output by 40% in order to meet the demands for food which the population rise will bring. European agriculture, for example, is currently increasing its output at an average rate of 2% per year as a result of improvements both in plant breeding and in agricultural practice. The explosion

Table 10.1 Acquisition of seed companies by some multinational corporations*

Multinational	*Seed company*
ICI	Garst, Sinclair
Shell	Nickerson
Sandoz	Northrup King, Rogers, Stauffer
Upjohn	Asgrow
Ciba–Geigy	Funk
Calgene	Stoneville
Pfizer	Trojan, Clemens
Unilever	Plant Breeding Institute

*Based on information in Kidd, 1987.

in the number of new biotechnology companies, and the acquisition of seed companies by multinationals in recent years (see Table 10.1) reflects the optimism with which the role of plant biotechnology in agriculture is perceived. There are presently approximately 500 companies and institutes worldwide involved in some aspect of crop improvement, and, as the field becomes increasingly competitive, with the commercial seed market being valued in 1987 at about $24 billion, the demands on biotechnological techniques to gain an edge over the competition will become heavier.

However, before the use of genetically-engineered crops becomes fully widespread, a number of interrelated ecological and political considerations have to be taken into account which could limit the commercialization of novel genotypes. It must be demonstrated that an engineered crop is not only able to grow and develop as well as (and hopefully better than) its untransformed relative, but that there are no associated risks to health and safety. In the United States there are now regulations, that can carry weight in a court of law, which govern the transport, handling and field trials of genetically-engineered organisms, although hybrid plants produced by somatic hybridization are excluded from this legislation. In Europe there are no generally agreed regulations or guidelines, each country making its own political decision. In the UK applications for field trials are made to the Advisory Committee on Genetic Manipulation, and if the appropriate experimental conditions are fulfilled, testing can go ahead. Rothamsted Experimental Station was first to release genetically-manipulated plants, potatoes transformed by *Agrobacterium rhizogenes*, into the field. In West Germany, however, there has been no permission granted for field testing, and similar restrictions hold in Denmark.

The safety risks from genetically-engineered plants themselves would be expected, in the vast majority of cases, to be negligible. The hypothetical dangers associated with the manipulation of plants can be broadly classified as (1) the accumulation of novel toxic compounds, (2) the acquisition of an uncontrolled, weedy growth habit, and (3) the transfer of, for example, resistance genes from an engineered crop to a weed by cross-pollination, conferring on the weed a selective advantage. Only specific genes with well-characterized gene products are likely to be transferred, however, and in this respect the result of recombination by genetic engineering is more predictable than that generated by conventional breeding. Furthermore, transformants would be thoroughly studied in the laboratory for alterations in growth habit and morphology before release into the field, and the possible accumulation of toxic metabolites in edible parts of the plant would certainly be tested before either introduction into a breeding programme or release for sale. The accidental transfer of resistance genes to weeds is only likely to be possible if the engineered and wild plant are closely related species, and if this should occur the maintenance of the gene in the population would require the appropriate selection pressure. Of course, resistance and a range of other genes are in any case currently introduced into crops by traditional breeding practices, and these are no less likely to be transferred to wild species than genes introduced by molecular biological techniques. Genetic engineering, at best, is no more than a useful addition to a plant breeding programme, and in itself is unlikely to generate unique risks. A bigger problem is to put the role, benefits and limitations of plant gene manipulation into perspective for the layman.

The issues concerning the release of genetically-engineered microorganisms are probably more contentious than are those for plants (Davis, 1987; Sharples, 1987), and in the United States this has resulted in $4.5 million being spent on risk assessment in 1988 under the control of the Environmental Protection Agency (EPA). The possible hazards are potentially the same, in general terms, as for plants (Marx, 1987): will the released organism survive?; will it proliferate?; will it spread beyond its intended area of application?; can it transfer its genetic material to other organisms?; and will the introduced genes confer toxic properties? The arguments in favour of release similarly point out that the engineered microorganisms are well characterized in the laboratory before release, but, unlike the situation with plants, the question about uncontrollable spread is less easily answered. Methods are being developed to follow the spread of specific strains in the environment, for example, by tagging them with marker genes, such as lactase, which can be assayed by supplying X-gal as substrate—the product is bright blue; other methods include the use of selectable markers such as resistance to unusual antibiotics (e.g. rifampicin), or probing for specific DNA sequences. There is also little information on how readily introduced genes will be transferred from engineered to other bacteria in normal soil conditions: there is evidence that this does occur, but the ecological impact is unknown. The biggest question mark concerns the possible toxic effects of introduced strains on the environment, not so much in relation to toxicity to humans, animals or plants, which is easy to determine prior to release, but to local populations of wild microorganisms, which may suffer from new competition. Nevertheless, it is unlikely that such possible effects are any more likely to take place if an introduced microorganism is genetically engineered than if it is not.

These arguments must be debated fully by companies, government bodies, independent scientists and the public; but it is only through a greater understanding of basic biology that sufficient facts will emerge to fuel the decision-making of politicians so that plant biotechnology can realize its full potential.

General Reading

Davis, B.D. (1987). 'Bacterial domestication: underlying assumptions', *Science* **235**, pp. 1329–1335.

Jain, H.K. (1988). 'Plant genetic resources and industry'. *Trends Biotechnol.* **6**, pp. 73–77.

Kidd, G.H. (1987). 'Macroeconomics of seed and plant biotechnology in the 1990s', in *Agricultural Economics and Technology. World Biotech. Rep.* **1**, pp. 1–8.

Marx, J.L. (1987). 'Assessing the risks of microbial release', *Science* **237**, pp. 1413–1417.

Sharples, F.E. (1987). 'Regulation of products from biotechnology', *Science* **235**, pp. 1329–1332.

Vasil, I.K. (1988). 'Progress in the regeneration and genetic manipulation of cereal crops', *Bio/Technol.* **6**, pp. 397–402.

Webb, C. and Mavituna, F. (Eds) (1987). *Plant and Animal Cells: Process Possibilities.* Chichester, Ellis Horwood.

Specific Reading

Hauptmann, R.M., Vasil, V., Ozias-Akins, P., Tabaeizadeh, Z., Rogers, S.G., Fraley, R.T., Horsch, R.B. and Vasil, I.K. (1988). 'Evaluation of selectable markers for obtaining stable transformants in the Gramineae', *Plant Physiol.* **86**, pp. 602–606.

Karuna-Karan, A. (1987). 'Product formulations from commercial-scale culture of microalgae', in *Agricultural Economics and Technology. World Biotech. Rep.* **1**, pp. 37–44.

Klee, H.J., Horsch, R.B., Hinchee, M.A., Hein, M.B. and Hoffman, N.L. (1987). 'The effects of overproduction of two *Agrobacterium tumefaciens* T-DNA auxin biosynthetic gene products in transgenic petunia plants', *Genes Devel.* **1**, 86–96.

Ooms, G. and Lenton, J.R. (1985). 'T-DNA genes to study plant development: precocious tuberisation and enhanced cytokinins in *A. tumefaciens* transformed potato', *Plant Mol. Biol.* **5**, pp. 205–212.

Rhodes, C.A., Lowe, K.S. and Ruby, K.L. (1988). 'Plant regeneration from protoplasts isolated from embryogenic maize cell cultures', *Bio/Technol.* **6**, pp. 56–60.

Schwartz-Sommer, Z. and Saedler, H. (1987). 'Can plant transposable elements generate novel regulatory systems?', *Mol. Gen. Genet.* **209**, pp. 207–209.

Shimamoto, K., Terada, R., Izawa, T. and Fujimoto, H. (1989). 'Fertile transgenic rice plants regenerated from transformed protoplasts', *Nature* **338**, pp. 274–276.

Glossary

aneuploid: Cell or organism possessing a chromosome number other than the haploid number or an exact multiple of it.

anti-sense RNA: RNA which is complementary to, and will bind to, sense or messenger RNA.

backcross: A sexual cross involving a hybrid and one of its parents.

callus: An unorganized mass of plant tissue, originally produced as wound tissue, but maintained on an artificial solidified nutrient medium.

CAP site: 5′ region of mRNAs characterized by the presence of methylated nucleosides, such as 7-methylguanosine.

cDNA (complementary DNA): DNA which is synthesized from (and is complementary to) an RNA template sequence, as catalysed by a reverse transcriptase.

C_0t value: A value measured as moles $litre^{-1}$ sec^{-1}) when a given proportion (usually 50%, $Cot\frac{1}{2}$) of dissociated strands of DNA in solution have reannealed. The $Cot\frac{1}{2}$ value depends in turn on the complexity of the DNA sequences: the higher the $Cot\frac{1}{2}$ value, the greater the DNA sequence complexity.

cultivar: A category of plants of the same species, but with a characteristic phenotype, generated through cultivation by man.

cybrid: Plant or cell which is a cytoplasmic hybrid produced by fusion of a protoplast and cytoplast (qv).

cytoplast: Enucleated protoplast.

enhancer: DNA sequence outside the coding region of a gene which acts to enhance the level of mRNA produciton. Commonly retains the effect if inserted in either a 5′-3′ or 3′-5′ orientation.

exon: Sequence within the coding region of a gene, the transcription product of which is found in a mature mRNA (cf. intron).

explant: Excised plant tissue used to initiate callus cultures or for regeneration or micropropagation purposes.

gene library (gene bank): Collection of DNA fragments (cDNA, genomic DNA) which have been cloned in a vector.

heterokaryon: Protoplast or cell with two, genetically distinct, nuclei, a product of somatic hybridization (qv).

homokaryon: Protoplast or cell with two, genetically identical nuclei, a product of somatic hybridization.

ideotype: Crop plant which possesses specific agronomic characteristics.

inbreeding: A breeding programme in which ancestrally related individuals are crossed.

In situ hybridization: Technique to localize either specific DNA sequences on a chromosome or specific mRNAs within a tissue by probing with a radio-labelled cDNA or RNA molecule.

intron: Sequence within the coding region of a gene which is transcribed but is subsequently excised (spliced) from the RNA. Mature mRNAs do not therefore contain sequences complementary to the intron DNA.

line breeding: A system of breeding in which a number of genotypes, which have each been selected for one or more characters, are crossed to develop a new variety.

linkage: The close physical association of parental alleles, such that they occur together in the offspring at a higher frequency than would be expected from the Mendelian segregation of each allele.

maturase: An enzyme which splices intron sequences from pre-mRNAs.

multigene family: A family of genes which encode structurally-related proteins, such as isozymes or related storage proteins.

northern blot/hybridization: The transfer and immobilization of RNAs to a membrane, to which complementary single-stranded DNA fragments can be hybridized.

outbreeding: The crossing of plants of different genotypes

pathovar: Variety of a species of pathogen.

phytoalexin: Chemical produced by some plants in response to pathogens.

polyadenylation site: Region of a gene, 3′ to the coding region, encoding the attachment of polyadenylic acid to the mRNA.

polygenes: Genes whose effects individually are too small to identify, but whose effects are additive in determining a quantitative trait.

polyploid: Cell or organism possessing a chromosome number which is an exact multiple, but more than twice that, of the haploid number.

presequence: A transient N-terminal extension of a polypeptide involved in the targeting of that polypeptide to a specific subcellular compartment (cf. transit peptide).

promoter: Region of a gene, 5′ to the coding region, to which RNA polymerase attaches to initiate transcription.

protoplast: Plant or fungal cell from which the wall has been removed, usually by enzymatic digestion.

restorer gene: Nuclear gene which can restore fertility to a male sterile line.

somaclonal variation: Genetic variation which is generated in cells cultured *in vitro*.

somatic embryo (embryoid): Bipolar structure of somatic cell origin which is capable of developing roots and shoots *in vitro*.

somatic hybridization: Hybridization which occurs as the fusion of somatic cell protoplasts.

Southern blot/hybridization: The transfer and immobilization of single-stranded DNA fragments to a membrane, to which specific complementary single-stranded DNA sequence or transcripts can be hybridized.

terminator: Region of a gene, 3′ to the coding region, which contains the signal sequences for the termination of transcription and for polyadenylation.

transit peptide: Transient sequence on a polypeptide, which may or may not be a presequence (qv), involved in the targeting of that polypeptide to a specific subcellular compartment.

transposon (also transposable element, jumping gene): DNA sequence which is able to excise from one locus and become inserted at a separate locus.

transposon mutagenesis: The mutagenesis of a gene by introducing into it a transposon (qv).

variety: A category of plants of the same species, but with a characteristic phenotype.

western blot/hybridization: The transfer and immobilization of proteins to a membrane, to which specific antibodies can be hybridized.

Index